NEW TOPICS IN CRYSTAL GROWTH RESEARCH

NEW TOPICS IN CRYSTAL GROWTH RESEARCH

GEORGE V. KARAS
EDITOR

Nova Science Publishers, Inc.
New York

For permission to use material from this book please contact us:
Telephone 631-231-7269; Fax 631-231-8175
Web Site: http://www.novapublishers.com

LIBRARY OF CONGRESS CATALOGING-IN-PUBLICATION DATA

New topics in crystal growth research / George V. Karas (editor).
p. cm.
Includes bibliographical references and index.
ISBN 1-59454-837-4
1. Crystal growth--Research. 2. Electronics--Materials. I. Karas, George V.
QD921.N486 2005
548'.5'072--dc22 2005014729

Published by Nova Science Publishers, Inc. ✦ New York

CONTENTS

Preface vii

Chapter 1 Optical Imaging and Control of Convection Around a Crystal Growing from its Aqueous Solution 1
K. Muralidhar, Atul Srivastava and P. K. Panigrahi

Chapter 2 Hydro/Solvothermal Route to Growth of Nanocrystals 85
Yuanfang Liu

Chapter 3 Growth of Alloy Nanowhiskers 125
Sin-iti Kitazawa

Chapter 4 The Growth of Low Resistance Metal Silicide Films by Solid State Reaction 153
J. Y. Feng, Y. He, J. F. Liu and R. N. Wang

Index 179

PREFACE

Experimental and theoretical aspects of crystal growth and its applications, e.g. in devices, are within the scope of this new book. Experimental and theoretical contributions are included in the following fields: theory of nucleation and growth, molecular kinetics and transport phenomena, crystallization in viscous media such as polymers and glasses; crystal growth of metals, minerals, semiconductors, superconductors, magnetics, inorganic, organic and biological substances in bulk or as thin films; molecular beam epitaxy, chemical vapor deposition, growth of III-V and II-VI and other semiconductors; characterization of single crystals by physical and chemical methods; apparatus, instrumentation and techniques for crystal growth, and purification methods; multilayer heterostructures and their characterisation with an emphasis on crystal growth and epitaxial aspects of electronic materials.

As reported in chapter 1, the importance of convection in determining crystal growth rates and crystal quality has been emphasized in the literature. However, measurement of convection patterns has posed certain difficulties. In this context, laser measurement techniques have become popular, though primarily as a flow visualization tool. The nature of convection and its relationship to solute distribution, crystal growth rates and quality remains to be fully established. For this reason, the dependence of the convection patterns on the process parameters has been explored in the present study. The process parameters varied are cooling rate of the solution and the rotation rate of the growing crystal. Crystal rotation is seen in this work as a means of diminishing the impact of buoyancy. It can also be an alternative to growth under microgravity conditions. The crystal size plays an influential role in determining the relative importance of buoyancy and rotation. An independent study of crystal size has also been presented. Mapping of convection patterns as well as the concentration field in the growth chamber is required to understand the mechanism of crystal growth and to establish appropriate conditions for growing large defect-free crystals. Unlike growth from melt and vapor, growth of a crystal from an aqueous solution is particularly amenable to optical visualization since the solution is transparent. It is possible to generate images of the convective field by exploiting changes in the refractive index that accompany changes in the density of the medium. The visualization technique is thus useful for on-line monitoring of the growth process. Optical techniques are ideal for mapping the properties of the solution during a crystal growth experiment because they are non-intrusive and inertia-free. In addition, the images recorded can be interpreted as path integrals in the direction of line-of-sight. Local properties can be subsequently reconstructed by using principles of tomography. In the present study, images from three refractive index-based methods namely, interferometry, schlieren and shadowgraph are presented. The suitability of these measurement techniques to image, analyze and interpret the convective field around a growing crystal is explored. The crystal material chosen for experiments is potassium-di-hydrogen-phosphate (KDP). The choice of KDP was based on the availability of refractive index and supersaturation data in the literature for image analysis. The three optical measurement techniques have been validated in the context of buoyancy-driven convection in a rectangular differentially heated cavity. This study is concerned with the imaging of buoyancy-driven convection in a fluid medium that is confined in a horizontal differentially heated rectangular cavity. The horizontal surfaces of the cavity provide a temperature difference, for initiating convection in the fluid. The vertical side walls are thermally insulated. Experiments have been conducted in a cavity of length 447 mm and 32 mm vertical height. The cavity is square in cross-section, and the imaging direction is parallel to its longer side. Convection in air and water have been investigated. Quantities of interest are the temperature profiles in the fluid medium and wall heat transfer rates. A direct comparison of the refractive-index based techniques shows that interferometry is best suited when density gradients in the fluid medium are small. For high density gradients, shadowgraph images contain the greatest amount of information. While interferograms are easy to analyze, it is difficult to retrieve concentration data from a shadowgraph. As a balance between ease of analysis and sensitivity to density gradients, schlieren emerges as the best option for crystal growth applications. On the basis of schlieren imaging, results obtained in the present work show that the growth rates are well-correlated with regions of large concentration gradients. Rotation serves to diminish concentration gradients near the crystal by homogenizing the solution. Accordingly, the growth rate is seen to be the highest in the buoyancy-driven convection regime, while the quality of the grown crystal is superior in the presence of rotation. The three dimensional distribution of concentration shows a strong correlation with the topography of the growing crystal.

Chapter 2 focuses on the hydro/solvothermal method of growing crystals has become widely accepted in the last few decades. The crystals of a variety of compounds and elements, such as simple oxides, complex oxides, chalcogenides, chalcohalides, metal elements, and non-metal elements, have been grown from hydro/solvothermal process. A few literatures and books have given reviews on the hydrothermal crystal growth of the above compounds and elements, mainly on their bulk crystal growth. In this account, we will focus on the nanocrystal growth of different types of compounds and elements under hydro/solvothermal condition. The nanostructures mainly include nanorods, nanotubes, nanoneedles, nanowires, nanospheres and nano- hollow spheres, nano-flakes, and nanodots. The influence factors, such as starting materials, solvents, pH, temperature, reaction time, on the growth of nanocrystals and on the morphology and size control of nanocrystals will be discussed since many physical and chemical properties of nanomaterials are depended on the shape and size to a large extent.

As reported in chapter 3, in recent years, whiskers have become the subject of intense study since their remarkable characteristics, such as ultrahigh strength, low-density, high-modulus, high-melting point. Whiskers are metallic and non-metallic filamentary single crystals that have micrometer-sized widths or diameters. Furthermore, a large interest on "nanowhisker" that have nanometer-sized widths or diameters have been attracted with the advance on the nanotechnology. Growth of nanowhisker has become a necessary and important phase of nanotechnology since the nanowhisker changed from a laboratory curiosity to a most desirable structural component. For the growth of whiskers, it has been proven by several technique which include stress-induced growth, evaporation-condensation, vapor-phase reaction, growth from the melt, electrolytic growth, growth from solution and directional solidification of binary eutectics. Those scale down methods have been applied for nanowhisker growths. For the growth of alloy nanowhiskers, it has not been reported in great numbers, since it is more complicate than that of a simple substance. In this paper, a novel growth of the alloy nanowhisker is described. The carbon - nickel - indium mixture thin films were prepared by pulsed laser deposition. The nanowhiskers were synthesized by the annealing of the films. The component of the alloy was analyzed to be Ni_2In_3, whose melting point is slightly lower than the annealing temperature. The unique growth mechanism of nanowhiskers is explained the supersaturated liquid Ni-In alloys were seeping out from a porous graphite thin film.

As reported in chapter 4, in recent years, low resistance metal silicides have been generally used in CMOS technology to decrease the sheet resistance of highly doped regions. In this article, we report the growth of three generations contact materials, i.e. C54-$TiSi_2$, $CoSi_2$ and NiSi, by solid state reaction in advanced CMOS technology. The research focuses on three aspects: the formation of C54-$TiSi_2$; the epitaxial growth of $CoSi_2$ on Si(100) substrates and the thermal stability of NiSi films. In Ti-Si system, a thin layer of Ta was introduced between Ti films and Si substrates in order to promote the formation of C54-$TiSi_2$. After rapid thermal annealing (**RTA**) at 580°C, the single C40-$TiSi_2$ phase was found in the sample with Ta interlayer, and C54-$TiSi_2$ phase was formed directly, bypassing the C49-$TiSi_2$ when annealed at higher temperature (650°C). The formation temperature of C54-$TiSi_2$ is lowered by 100°C compared with the sample without the Ta layer. In Co-Si system, highly textured Co films were obtained by heating the Si substrates to an adequate temperature during the deposition of cobalt. The direct solid phase reaction between textured Co films and Si(100) substrates at high temperatures produced epitaxial $CoSi_2$ with good quality. Rutherford backscattering/channeling spectrometry and high resolution transmission electron microscopy were performed to evaluate the epitaxial quality of the $CoSi_2$ films. A good channeling χ_{min} value of 20% was measured for the $CoSi_2$ films formed after rapid thermal annealing at 850°C, which is comparable to that of the $CoSi_2$ films obtained by Ti-interlayer mediated epitaxy (TIME). Similar with TIME, the intermediate phase of Co_2Si was skipped in this case. The concept of "thermodynamic barrier" has been employed to explain the improvement in the epitaxy of $CoSi_2$ films compared with the reaction between amorphous Co films and Si substrates. In Ni-Si system, a thin Pt interlayer was used to raise the phase transition temperature by about 100°C and thereby greatly improve the thermal stability of NiSi films. It was also found that the thermal stability of NiSi films improves with the decrease of the film thickness in Ni/Pt/Si system, implying a promising future for the application of NiSi films in shallow junction contacts. A semi-quantitative analysis was carried out to clarify the effect of Pt on the activation energy for nucleation and critical nucleus size of $NiSi_2$, which explains the above the experimental results.

In: New Topics in Crystal Growth Research
Editor: George V. Karas, pp. 1-83
ISBN 1-59454-837-4

Chapter 1

OPTICAL IMAGING AND CONTROL OF CONVECTION AROUND A CRYSTAL GROWING FROM ITS AQUEOUS SOLUTION

K. Muralidhar, Atul Srivastava and P. K. Panigrahi
Department of Mechanical Engineering
Indian Institute of Technology Kanpur
Kanpur 208 016 UP (India)

ABSTRACT

The importance of convection in determining crystal growth rates and crystal quality has been emphasized in the literature. However, measurement of convection patterns has posed certain difficulties. In this context, laser measurement techniques have become popular, though primarily as a flow visualization tool. The nature of convection and its relationship to solute distribution, crystal growth rates and quality remains to be fully established. For this reason, the dependence of the convection patterns on the process parameters has been explored in the present study. The process parameters varied are cooling rate of the solution and the rotation rate of the growing crystal. Crystal rotation is seen in this work as a means of diminishing the impact of buoyancy. It can also be an alternative to growth under microgravity conditions. The crystal size plays an influential role in determining the relative importance of buoyancy and rotation. An independent study of crystal size has also been presented.

Mapping of convection patterns as well as the concentration field in the growth chamber is required to understand the mechanism of crystal growth and to establish appropriate conditions for growing large defect-free crystals. Unlike growth from melt and vapor, growth of a crystal from an aqueous solution is particularly amenable to optical visualization since the solution is transparent. It is possible to generate images of the convective field by exploiting changes in the refractive index that accompany changes in the density of the medium. The visualization technique is thus useful for on-line monitoring of the growth process.

Optical techniques are ideal for mapping the properties of the solution during a crystal growth experiment because they are non-intrusive and inertia-free. In addition, the images recorded can be interpreted as path integrals in the direction of line-of-sight.

Local properties can be subsequently reconstructed by using principles of tomography. In the present study, images from three refractive index-based methods namely, interferometry, schlieren and shadowgraph are presented. The suitability of these measurement techniques to image, analyze and interpret the convective field around a growing crystal is explored. The crystal material chosen for experiments is potassium-di-hydrogen-phosphate (KDP). The choice of KDP was based on the availability of refractive index and supersaturation data in the literature for image analysis.

The three optical measurement techniques have been validated in the context of buoyancy-driven convection in a rectangular differentially heated cavity. This study is concerned with the imaging of buoyancy-driven convection in a fluid medium that is confined in a horizontal differentially heated rectangular cavity. The horizontal surfaces of the cavity provide a temperature difference, for initiating convection in the fluid. The vertical side walls are thermally insulated. Experiments have been conducted in a cavity of length 447 mm and 32 mm vertical height. The cavity is square in cross-section, and the imaging direction is parallel to its longer side. Convection in air and water have been investigated. Quantities of interest are the temperature profiles in the fluid medium and wall heat transfer rates.

A direct comparison of the refractive-index based techniques shows that interferometry is best suited when density gradients in the fluid medium are small. For high density gradients, shadowgraph images contain the greatest amount of information. While interferograms are easy to analyze, it is difficult to retrieve concentration data from a shadowgraph. As a balance between ease of analysis and sensitivity to density gradients, schlieren emerges as the best option for crystal growth applications.

On the basis of schlieren imaging, results obtained in the present work show that the growth rates are well-correlated with regions of large concentration gradients. Rotation serves to diminish concentration gradients near the crystal by homogenizing the solution. Accordingly, the growth rate is seen to be the highest in the buoyancy-driven convection regime, while the quality of the grown crystal is superior in the presence of rotation. The three dimensional distribution of concentration shows a strong correlation with the topography of the growing crystal.

Key words: Convection, Supersaturated solution, Ramp rate, Rotation, Imaging, Interferometry, Schlieren, Shadowgraph, Tomography, Crystal quality.

1. INTRODUCTION

Large crystals with a high degree of perfection are required in a variety of applications. Optical crystals of good quality find utility in critical technology areas such as high power lasers, higher harmonic generation and in nuclear fusion. A class of such crystals can be grown from their supersaturated solution in water, during a slow cooling process. Growth of such crystals from an aqueous solution is one of the commonly used techniques in the industry. The technique can be used for growing optical crystals such as KDP and proteins such as lysozyme. A crystal growing from its aqueous solution creates a three-dimensional solute distribution in its vicinity. The solutal concentration gradients and hence the gradients in the density of the solution are responsible for the evolution of buoyancy-driven convection currents in the growth chamber. The buoyant convection currents influence the magnitude of the concentration gradients prevailing along the growth surfaces. In turn these control the stability of the growth process and the overall crystal quality. The concentration gradients in the growth chamber are significantly altered when the crystal is given rotation. Rotation can hence be viewed as a method of controlling convection during the growth process.

Low temperature solution growth methods are applicable to materials that have moderate to high solubility in temperatures up to100^oC at atmospheric pressure. A number of organic and inorganic materials fall in this category and can be crystallized using this technique. The advantages here include relatively low temperature handling of the equipment and a good degree of temperature control to within $\pm 0.01^oC$. The grown crystals show full natural morphology. In addition, online studies of surface growth features and growth rate studies of different faces are possible. Since temperature gradients involved in a given process are low, thermally generated strains in the grown crystal are small. The disadvantage of low temperature solution growth is the slow growth rates (0.1 to 10 mm per day). In many cases the ease of solvent inclusion into the growing crystal is a limiting factor. Since many crystals are grown from water as the solvent, the grown crystal is hygroscopic in nature; this limits its use to applications in which water molecules are excluded in the lattice. A large fraction of the crystals produced from low temperature solutions are grown by using water as a solvent owing to its high solvent action, chemical stability, low viscosity, low toxicity and low cost. Other solvents include ethanol, methanol, acetone, carbon tetrachloride, hexane, and xylene. Examples of technologically important crystals grown from a low temperature aqueous solution include potassium di-hydrogen phosphate (KDP), potassium di-deuterium phosphate (DKDP), tri-glycine sulphate (TGS), potassium acid phthalate (KAP), lithium arginine phosphate (LAP), and urea. Recently, low temperature solution growth method has found its applications towards the growth of protein crystals such as lysozymes.

A crystal growing from its aqueous solution creates a three-dimensional solute distribution in its vicinity. The mechanism involved is the withdrawal of the solute from the solution, causing the solute concentration to decline in the neighboring solution. The solutal concentration gradients, and hence the gradients in the density of the solution are responsible for the evolution of buoyancy-driven convection currents in the growth chamber [1]. The convection currents are the main drivers for the solute from the solution rich in salt as it is transported to the surface of the growing crystal. In the absence of convection, solute transport is mainly by molecular diffusion. Since diffusion coefficients in liquids are small, small fluid velocities greatly increase mass transfer (over the diffusion values) and hence the growth rate of the crystal. It is understandable that changes in the symmetry pattern of the flow field and unsteadiness can lead to a lowering of the crystal quality and the growth rate. Hence an optimum strength of convection currents is desirable to maintain a balance between the quality of the growing crystal and its growth rate. The strength and orientation of buoyancy driven convection currents are intricately linked with the size and morphology of the growing crystal. These in turn depend upon the process parameters namely the supersaturation level of the solution and the rate at which it is cooled. More recently, convection has been found to be quality-limiting mechanism in the growth of protein crystals from their solution. Here, convection has the capability of distorting the effects of mechanisms such as surface tension and magnetic fields and thus influences the crystal growth process.

Buoyancy-driven convection currents influence the magnitude of the concentration gradients prevailing along the growth interfaces. In turn, the gradients control the stability of the growth process and the overall crystal quality. It has been experimentally noted that growth in free convection regime is often limited by density stratification in the vicinity of the growing crystal leading subsequently to unwanted nucleation of solute in the growth chamber. The concentration gradients in the growth chamber are significantly altered when the crystal is given a rotation. An optimum rotation rate tends to stabilize the perturbations along the rotational axis eventually leading to an axisymmetric concentration distribution over the crystal. The stirring of the solution also reduces the natural convection-induced temperature oscillations by homogenizing the bulk solution. Rotation can hence be viewed as a method for controlling convection during the growth process by diminishing the impact of buoyancy.

To ensure the growth of high-quality large crystals, it is important to understand from a fundamental point-of-view, the transport phenomena involved during solute deposition from the solution to the crystal surfaces. Simultaneously, visualization techniques are required to monitor the crystal growth process itself during its progress. Unlike growth from melt and vapor, growth from an aqueous solution is particularly amenable to flow visualization, since it is transparent. It is possible to generate images of the convective field by exploiting changes in the refractive index that accompany changes in the density of the medium. Optical visualization techniques are thus useful for online monitoring of the growth process.

Optical techniques are ideal for mapping the properties of the solution during a crystal growth experiment. Since a probe need not be introduced in the field of study, they are non-intrusive and practically inertia free. Over the past decade, laser measurement techniques have become popular, though primarily as a flow visualization tool. Recent work however has emphasized the possibility of quantitative measurements as well. A majority of optical techniques are field techniques in the sense that an entire cross-section of the physical region can be mapped. They require the medium to be transparent, and are thus suitable for the measurements in liquids. Optical methods that utilize the dependence of refractive index of light on density (and indirectly on concentration and temperature) can be configured in many different ways. Three available routes are:

1. Interferometry, where the image formation is related to changes in the refractive index with respect to a reference environment,
2. Schlieren, where light deflection in a variable refractive-index field is captured, and
3. Shadowgraph, where the reduction in light intensity on beam divergence is employed.

In the context of crystal growth from an aqueous solution, a unique relationship can be established between the refractive index and the local density of the medium under study. Under normal process conditions, the solution is practically incompressible and the density does not depend on pressure. At any time instant, the solution is also at a spatially uniform temperature. Hence, density changes correlate with those in concentration alone, and the three methods become applicable for concentration field measurement in the fluid medium. The three techniques referred above yield the path integral of the density field in the direction of line-of-sight. The distribution of density in three dimensions can then be extracted from the recorded images using principles of tomography.

The present article reports measurement of the solute concentration distribution around a KDP crystal growing from its aqueous solution under slow cooling conditions. The choice of KDP (molecular mass 136.09) was based on the availability of refractive index and supersaturation data in the literature for image analysis. The solution adjacent to the crystal is depleted of the salt and is close to the saturated state, while it is supersaturated in the far-field. The difference in solute concentrations drives a diffusive mass flux in the initial stages of growth, while the flux is controlled by fluid convection at later times (Figure 1). In the context of crystal growth, the present article contains discussions on refractive-index methods, their validation in a buoyancy-driven flow experiment, comparison of interferograms, schlieren and shadowgraph for crystals growing under nominally identical experimental conditions, schlieren imaging and analysis of convection patterns, possibility of control by rotation, and tomographic reconstruction of the concentration field around the growing crystal. The physical problem taken up for validation and comparison of the three optical methods is buoyancy-driven convection in a fluid medium confined in a rectangular cavity (Gebhart *et al.* [2]). The fluid is heated from below, cooled from the top and the cavity has insulating side walls. The flow pattern associated with this configuration shows a sequence of transitions from steady laminar to unsteady turbulent flow.

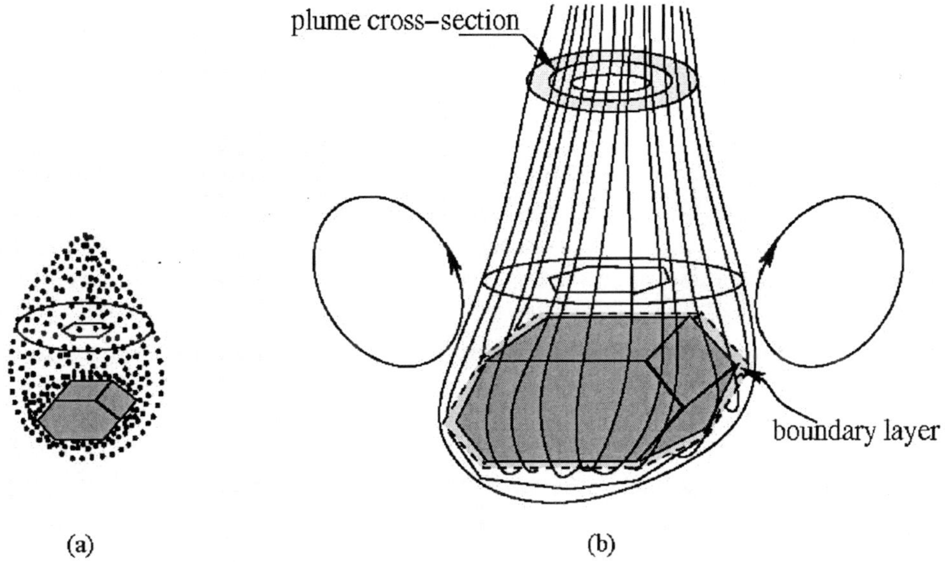

Figure 1. Schematic drawing of the concentration distribution around a growing crystal. (a) Diffusion dominated growth shown by dots, (b) Growth in stable convection regime in the presence of a buoyant plume. Arrows indicate fluid motion in the bulk of the solution.

2. Literature Review

The following sections present a survey of the published literature on imaging fluid convection using optical techniques. The review covers imaging and characterization of the convective fields induced due to thermal gradients, solute concentration gradients and measurements of three-dimensional temperature and concentration fields using tomography algorithms. The emphasis is on the survey of published literature related to the imaging of buoyancy-driven convection during crystal growth processes from an aqueous solution.

2.1. Imaging of Heated and Thermally Buoyant Flows

Buoyancy-driven flow and heat transfer are performance-limiting steps in several engineering processes. Examples can be seen in applications ranging from cooling of electronic components, growth of semiconductor crystals to material processing. Lewis *et al.* [3] conducted a quantitative study of temperature distribution using shadowgraph. The focus was on the determination of the temperature field in a laser ignition experiment wherein an ethylene-air mixture was heated by a high power laser, and the hot gas was illuminated with a collimated laser beam. To cross-check the results obtained from the shadowgraph experiments, the authors determined the temperature field under similar test conditions using Rayleigh light scattering and demonstrated a good agreement between the results obtained from these two approaches. Kosugi *et al.* [4] applied the schlieren technique to experimentally record gas temperature profiles in the shock region of excimer lasers and correlated them to the xenon concentration in helium gas. Muralidhar *et al.* [5] studied transient natural convection in a square cavity in the intermediate Rayleigh number range with air as the working fluid using a Mach-Zehnder interferometer. Chandrasekhara *et al.* [6] discussed a high-speed phase locked interferometry system designed and developed for real-time measurements of dynamic stall for flow over a pitching aerofoil. Zhong and Squires [7] used interferometry to evaluate organized structures in high-speed flow past a circular cylinder. The authors reported the similarities between compressible and incompressible wakes as well as similarities in the turbulent structures. Dupont *et al.* [8] discussed the use of electronic speckle interferometry for visualizing isotherms as well as streamlines in a Rayleigh-Benard configuration. Schopf *et al.* [9] adopted a shadowgraph approach for the study of convective flow in a water-filled square cavity which was differentially heated and cooled from the opposing vertical sidewalls. Tanda and Devia [10] performed a quantitative schlieren study of free convective heat transfer from heated plates and heat transfer from vertical parallel channels to the surrounding fluid. Howle *et al.* [11] carried out flow visualization of convective flow in a porous medium using the shadowgraph method. Agrawal *et al.* [12] described a color schlieren approach to temperature field measurements in gas jets using white light. Dezhong and Tiange [13] applied real time laser interferometric tomography for the measurement of three-dimensional temperature fields generated by two heated rods.

A review of the literature shows that laser measurement techniques that depend on refractive index changes in the fluid medium are regularly used in engineering research. Extension of these techniques to map three dimensional properties of the flow field and flow unsteadiness is to be seen in specialized applications.

2.2. Three-Dimensional Reconstruction of Density Fields

Recent literature on the application of refractive index based methods to the study of three-dimensional flow and thermal fields is presented here. Snyder and Hesselink [14] carried out instantaneous optical tomographic reconstruction of non-stationary fluid flow using holographic interferometry. The authors determined the spatial distribution of helium concentration in a co-flowing jet of helium in air. Watt and Vest [15] studied structures of turbulent helium jets in air by recording path integral images based on the refractive index variation with a phase-shifted interferometer. Subsequently, the authors reconstructed the helium concentration field over the jet cross-section. Michael and Yang [16] discussed measurement of the three-dimensional temperature field in Rayleigh-Benard convection. The authors used a Mach-Zehnder interferometer, while reconstruction was performed using an iterative algebraic reconstruction technique. McMackin and Hugo [17] employed high-speed optical tomography system for imaging dynamic transparent media. The system was used to obtain spatially resolved high-speed cross-sectional temperature images of a round heated jet. Agrawal *et al.* [12] employed quantitative rainbow schlieren deflectometry with tomography for measurements of temperature in three-dimensional gas flows. The beam deflection data acquired from the schlieren images at multiple viewing angles were used to reconstruct refractive index data and temperature distributions with a modified convolution back projection (CBP) tomographic algorithm. Ko and Kihm [18] applied laser speckle photographic tomography to reconstruct the concentration field generated by asymmetric single and double helium jets in air. Mishra *et al.* [19] conducted an interferometric study of Rayleigh-Benard convection in air at intermediate Rayleigh numbers using tomography with limited projection data. More recently, Mishra *et al.* [20] carried out a detailed evaluation of iterative tomography algorithms for convection in a cylindrical cavity. The authors addressed the problem of handling incomplete and limited projection data that is frequently encountered in the engineering context.

In the above discussion, holography as a technique for determining the fluid density in three dimensions has not been included. In the opinion of the authors, laser imaging followed by tomography is a more convenient route for studying three dimensionality in flow and thermal fields.

2.3. Laser Imaging of Convection Around a Growing Crystal

Optical techniques have been employed to image the convective field around a growing crystal from its aqueous solution and quite a few studies have been reported. These experiments are described below in some detail.

In an early paper, Chen *et al.* [21] studied convection phenomena during the growth of sodium chlorate crystals from their aqueous solution using time-lapse schlieren cinematography. The authors addressed three major issues in their study: 1) Determination of conditions for transition from stable to unstable convection as a function of supersaturation of the solution and crystal size; 2) Characterization of plume width, its unsteadiness, and eddy frequency; 3) Effect of convection on the growth rate of the crystal. In addition, the authors studied the effect of mechanical shock on stable convection in the vicinity of the transition point to instability in an attempt to measure the strength of the convection currents. Grashof number based on solutal concentration was used to define the index of stability of the convective phenomena and an empirical relation was established between its critical value (at which transition from stable to unstable convective motion takes place) and the crystal height. The study also showed a good correlation between the strength of convection currents with plume characteristics such as its width and unsteady movement. Average growth rate of the crystal was measured by weighing as well as sizing the crystals before and after each experiment. The mass transfer rate was reported in terms of Sherwood number, a non-dimensional parameter. The authors also fitted an empirical relationship between Grashof number and the average dimensionless concentration gradient at the growth interface of the crystal. In a separate study with sodium chlorate growing from its aqueous solution, Chen *et al.* [22] showed that with increasing supersaturation and horizontal crystal dimensions, the behavior of the convective plume changed from steady laminar flow to oscillatory flow with periodic shedding of eddies in the upward direction. The authors reported an improved correlation between Sherwood number and Rayleigh number. The latter is a parameter (like Grashof number) that governs the strength of buoyancy-induced convection currents in the growth chamber. The study showed that the growth rate was slightly higher when the convective plume was laminar than when it was irregular. The horizontal growth rate of the crystal was shown to be greater than the vertical growth rate, with the ratio tending to decrease with increasing plume instability.

In one of the earliest studies on protein crystals, Shlichta [23] reviewed the feasibility of using optical techniques for mapping convection, temperature and solute concentration in the solution around a growing protein crystal. Techniques considered were schlieren, interferometry and holography. In terms of sensitivity, the study showed that schlieren (based on the knife-edge principle) and interferometric techniques are marginally sensitive whereas phase-contrast schlieren, ultraviolet solute absorption, and laser Doppler anemometry are capable of mapping the velocity field in the crystal vicinity. In addition, author proposed an optical technique based on absorption-interferometry for simultaneous mapping of concentration and temperature in the solution around a growing crystal. Double-exposure holography and birefringence (measured as changes of the interference color of the crystal mounted between crossed polarizers) were shown to be suited for online measurement of growth rates of the protein crystal from its aqueous solution.

Bedarida [24] employed holographic interferometry (one beam and multi-directional) for the study of crystal growth from an aqueous solution. The mean value of the concentration field averaged over the width of the solution was obtained and the growth rates of the crystal faces were estimated. Three-dimensional information was derived from the multi-directional holographic technique by reconstructing solute concentration on planes parallel to the crystal face.

Bredikhin *et al.* [25] experimentally studied growth kinetics of KDP crystals growing from an aqueous solution in the free convection regime. Movement of the crystal faces with time was measured by a laser interference-polarization technique. The growth rates of the crystal faces in free convection regime were compared to those in forced convection (achieved by pumping solution in the growth chamber). It was shown that a significantly higher growth rate can be achieved in the forced convection regime. In addition, the authors also studied the effect of impurities such as Al^{3+} and Fe^{3+} on the crystal growth rate.

The role of buoyancy-driven convection was emphasized by Onuma *et al.* [26] in a case study of barium nitrate crystal growing from its aqueous solution. The authors employed schlieren technique and Mach-Zehnder interferometry for visualizing the convective field around the growing crystal. The effects of diffusion boundary layer and buoyancy driven convection upon the growth kinetics of crystals were investigated in relation to the bulk supersaturation of the solution. Schlieren technique using white light was used for qualitative observation of the growth process. An interferometer was used to measure quantitatively the gradient of the refractive index within the diffusion boundary layer. The authors classified convection patterns into three regimes: diffusion dominated, unstable movement of weak convective plume, and stable convection plume, depending upon the level of supersaturation of the solution. The study demonstrated the effect of strength and orientation of buoyancy-driven convection currents on the growth rates of the individual faces of the crystal. The top face was seen to grow faster when the convection plume moved laterally and a higher growth rate of the side faces was observed in the presence of a stable convective plume. In a later work [27], the authors studied the effect of buoyancy-driven convection upon the surface morphologies (micro-topography) of different faces of $Ba(NO_3)_2$ and CdI_2 crystals. Schlieren technique was employed to visualize the buoyancy-driven convection currents. Onuma *et al.* [28] employed Mach-Zehnder interferometer for simultaneous measurements of the growth rate and surface super-saturation of K-alum crystals growing in an aqueous solution with varying flow velocities. The authors demonstrated that the two-dimensional profile of supersaturation levels varies significantly with flow velocity.

Bedarida *et al.* [29] illustrated the use of diffraction techniques based on the principle of phase measurements to study concentration and temperature gradients in aqueous solutions. As a case study, sodium chlorate crystals were used. The authors demonstrated that the technique is particularly useful for making local measurements in the solution over a small volume (whose dimensions are of the order of the wavelength of light) of the solution. In view of its simplicity and low electric power requirements the technique is useful for microgravity applications in outer space.

A Mach-Zehnder interferometer was also employed by Mantani *et al.* [30] for electronic measurement of concentration gradient around a KDP crystal growing from its aqueous solution. The authors established a linear relationship between the refractive index gradient and the growth rate of the crystal. Onuma *et al.* [31] constructed a microscopic Mach-Zehnder type real time phase shift interferometer for the measurement of concentration distribution around a growing or dissolving crystal in free convection regime. Sodium chlorate ($NaClO_3$) crystal was used to measure the surface concentration under different conditions of convection. It included relatively steady as well as fluctuating concentration fields. The study showed that the phase-shift interferometer has around 25 times higher sensitivity compared to the conventional two-beam interferometer. It also has higher time resolution to capture the fluctuations in the concentration field. Using Abel transformation, the authors obtained the actual surface concentration distribution of the growing crystal from the quasi-three-dimensional concentration profile. Sunagawa *et al.* [32] investigated transport phenomena during the growth and dissolution of barium nitrate ($Ba(NO_3)_2$), and K-alum crystals, their effects on the growth kinetics and degree of perfection of the grown crystals. The authors employed monochrome schlieren technique for visualizing convection as well as the diffusion boundary layer around growing and dissolving crystals. For quantitative measurements of concentration variation Mach-Zehnder interferometry was adopted. For *in-situ* observations of surface micro-topographic features and growth rate measurements of the growing crystal the authors used an optical microscope as well as a Michelson interferometer. The authors recommended the use of polarization microscope for *in-situ* identification of dislocation characteristics of individual growth hillocks (growth centers). The study emphasized the importance of solution flow in the growth chamber for achieving uniform supersaturation level over a wider surface area of the growing crystal by minimizing the effect of the buoyant plume.

Yu *et al.* [33] proposed holographic phase-contrast interferometric micro-topography (HPCIM), phase conjugation interferometry (PCI) and laser diffractometry as measurement techniques for studying real-time crystal growth and electrode processes. KDP was taken as the sample material for conducting experiments. The authors demonstrated the use of the HPCIM technique for 1) quantitative measurements within the boundary-layer and interface patterns during the crystal growth process, 2) precise measurement of saturation for controlling supersaturation level of a given solution, and 3) studying phenomena taking place at the electrode-electrolyte interface. Phase conjugation interferometry (PCI) was recommended over the conventional Michelson interferometry for the characterization of surface morphology such as slope of the growth hillock and step heights of the growing crystal. For precise measurement of crystal growth rates, the authors proposed a laser diffraction metering technique (LDMT) for low-temperature solutions, high-temperature solutions and vapor-phase crystal growth processes.

Kleine *et al.* [34] carried out an experimental study of diffusion around a growing crystal from its aqueous solution using dark field-type schlieren microscope. The viability of the method was demonstrated by the *in situ* measurement of concentration distribution in the diffusion field around silver dendrites. The dendrites were growing from an aqueous $AgNO_3$ solution by electrolysis. The microscope enabled the calculation of solute iso-concentration lines around the growing crystals. The technique was shown to be rugged and relatively insensitive to vibrations compared to interference techniques. The drawback related to the fact that the gradients in concentration along x and y directions could not be simultaneously recorded. Hence it was not suitable for applications in rapidly changing and complex diffusion fields. Kransinski [35] conducted experiments to study the evolution of the diffusion field around a KDP crystal growing in gel. The optical technique employed for visualization of concentration distribution in gel was speckle interferometry. The linear kinetic coefficient for the KDP crystal growing in gel was calculated from the shape of concentration profiles and its value was shown to be significantly lower than that obtained in the case of solution growth. The author compared the results with the simulation based on the one-dimensional diffusion model. Bredikhin *et al.* [36] proposed a technique based on the optical schlieren method for *in situ* investigation of the morphology of a large growing KDP crystal (8×8 cm^2 cross-section) under rapid growth conditions and for monitoring the crystal growth process to provide quality control. The surface morphology of the crystal was characterized in terms of inclination angles of growth hillocks (measured with a sensitivity of 2' using the proposed technique) on the crystal surface. The authors showed that when the growth rate of the crystal exceeds an optimal value, a wavy structure appears on the crystal surface. The waviness can lead to the origin of micro-inclusions and formation of inter-layers in the crystal. The proposed optical technique was shown to be useful towards controlling the morphology of the growing face by optimizing experimental conditions.

Inoue *et al.* [37] studied the effect of supersaturation and super cooling of the aqueous solution on the growth faces of $Na_2S_2O_3.5H_2O$ crystals. The authors employed the schlieren method and the Mach-Zehnder interferometer for *in situ* observation of the formation process of spherulites. It was shown that the growth shapes changed from prismatic (normal growth) to spherulite (abnormal growth) with an increase of supersaturation level of the solution. Piano *et al.* [38] demonstrated the use of electronic speckle interferometry (ESPI) for investigating convective phenomena around a growing KDP crystal. It was used for the measurement of concentration gradients as well as growth velocities of the crystal faces. The authors showed the potential of the technique for studying diffusion phenomena by capturing the refractive index gradients at the sugar-water interface in a solution of sugar in water. In a separate study [39], the authors performed dissolution and growth experiments for KDP crystals and observed convection using electronic speckle interferometry. The authors showed the presence of three regions of different densities around the dissolving/growing crystal that were independent of the geometry of the cell. The potential of ESPI was also demonstrated for calculating the temporal and spatial distribution of the solute concentration around the growing crystal.

Duan and Shu [40] applied phase shift interferometry for the study of the kinetics of sodium chlorate ($NaClO_3$) crystals. Both crystallization and dissolution phase of the crystal in its aqueous solution were considered. The optical technique was employed to visualize the convective plume in the crystal vicinity and for *in situ* measurement of the distribution of refractive index gradient. The sensitivity of the optical diagnostic system was shown to be higher than that of the conventional two-beam interferometry. In addition to the visualization of convective phenomena, the potential of phase-shifting interferometry for *in situ* imaging and characterization of interfacial morphology of a growing KDP crystal was demonstrated by Booth *et al.* [41]. The method was applied to monitor the (101) face of the crystal growing from its aqueous solution. The formation and evolution of solution-flow induced step bunches were imaged using the optical technique. In a separate application of real-time phase-shifting interferometry, Maruyama *et al.* [42] observed transient double diffusion fields around a rapidly growing sodium chlorate crystal from its aqueous solution. A two-dimensional transient numerical code was also developed to simulate the diffusion field around the crystal and an accurate measurement of solutal gradients in the crystal vicinity was carried out using the computer code. The authors estimated the growth rate of the crystal by combining experimental data with results of numerical simulation. Bredikhin *et al.* [43] employed the schlieren method with some improvements (over the set up as used in [36]) to achieve a higher optical resolution. Competing growth centers and step bunching in KDP crystal growth from its aqueous solution were investigated. The study focused on the morphology of (100) and (101) crystal faces when fed by impingement controlled solution flow. Schlieren observations showed that the brightness difference between the neighboring slopes of the hillocks is proportional to the angle between them and hence can be used as the basis for characterizing the surface morphology of the growing crystal face. The authors concluded that the mechanisms of step bunching and competition between growth centers are essential for good crystal quality. They are quite affected by the solution flow. When the flow is weak, multi-center growth occurs and an increase in flow intensity results in single-center growth of the crystal face.

In a recent study, Srivastava *et al.* [44] compared interferometry, schlieren and shadowgraph for studying buoyancy-driven convection patterns around a growing KDP crystal from its aqueous solution. While the spatial distribution of concentration differences in the vicinity of the crystal was revealed by all the techniques, it was concluded that the laser schlieren technique is best suited for process control. The basis of this observation was ease of instrumentation, image clarity and simplicity of data analysis. The study also brought out the limitation of free convection crystal growth experiments in the form of density stratification around the growing crystal. In a subsequent work [45], the authors studied the effect of ramp rate and crystal rotation on the convective field around a growing KDP crystal. Laser schlieren technique was employed as a diagnostic optical tool to map the concentration field around the crystal. It was shown that buoyancy-driven convection intensifies with increasing ramp rate as well as with increasing size of the crystal. Rotation tends to diminish the concentration gradients and improves the symmetry and uniformity of concentration distribution over the crystal faces. The findings of [44] and [45] are discussed at length in Section 9 of this article.

2.3.1. Protein Crystals

Optical techniques are presently being applied for visualization and control of buoyancy-driven convection in the growth of protein crystals. Convection currents have an adverse impact on the quality of the growing protein crystals. Steps have to be taken to suppress the buoyancy-induced flow during the growth process. Miyashita *et al.* [46] made use of a two-beam Michelson interferometer for *in situ* observation of solute concentration distribution around a growing lysozyme crystal from its aqueous solution. From the experimentally recorded interferograms, concentration profiles around the crystal, growth rates and diffusion coefficient of the lysozyme molecules were estimated. Vekilov and Rosenberger [47] experimentally studied the effect of solution flow on the growth kinetics and interface morphology of protein lysozyme crystals growing from their aqueous solution using laser interferometry. The authors showed that continuous circulation of the solution at an optimum rate provides an alternative to microgravity. Higher kinetic stability for the solute transport is achieved that in turn results in higher crystal quality. Hou *et al.* [48] employed the two-beam Michelson interferometer to visualize concentration gradient layers around tetragonal lysozyme crystals growing from gel. Gel was used as a solvent in order to suppress convection and mimic the microgravity environment. The surface concentration properties (its width and gradients) around the crystal as measured in gel were found to be noticeably different from that measured in an aqueous solution. In particular, concentration gradients were wider and transport rates were slower for gel-grown protein crystals. Yin *et al.* [49] reported an interferometric study of concentration measurement in a protein solution under the influence of a strong magnetic field. The growth process of tetragonal crystal of lysozyme was investigated using a Mach-Zehnder interferometer. The study showed that the crystal growth process under the influence of a magnetic field was slower when compared to that without a magnetic field. However, there was no significant effect of varying the magnitude of the magnetic field on the buoyancy-driven convection currents in the lysozyme solution. Sazaki *et al.* [50] employed two-beam Michelson interferometry for determining the solubility curve of protein lysozyme crystals. The equilibrium temperature of the crystals in solution was estimated on the basis of the bending of interference fringes at a given protein concentration. The study showed that the technique is insensitive to the impurities in the solution. It is comparatively less time consuming while compared to other conventional methods used for solubility measurements. Kawaji *et al.* [51] employed two different techniques, namely, photo chromic dye activation and particle tracking velocimetry, for investigating the occurrence of Marangoni and natural convection during the growth of lysozyme protein crystals. Marangoni convection is related to gradients in the interfacial tension, while natural convection arises from density gradients in the bulk of the solution. The visualization technique of the authors made use of a photo chromic dye material with a UV light absorption spectrum that changes its molecular structure upon UV light irradiation. The photo chromic reaction changes the liquid from clear to visible colors, resulting in the formation of visible dye traces. The velocity of the liquid containing the photo chromic dye can be obtained from the displacement of the dye traces over a given interval of time. In particle tracking velocimetry, small tracer particles are dispersed in the fluid and are illuminated with a visible laser light sheet. Flow velocities can be estimated by determining the displacement of the tracer particles with time. The fluid velocity estimates made by the two techniques indicated that the protein crystal growing from an aqueous solution is highly resistant to Marangoni convection. Ramachandran and Leslie [52] have recently shown that magnetic fields may be useful in mimicking the microgravity environment of space and can be employed for controlling the strength of convection currents during the growth of protein crystals. The authors analyzed the physical phenomena using a mathematical model and validated the results against the experimental observations.

The survey of articles above clearly shows that the importance of fluid convection in crystal growth from an aqueous solution is well-recognized. The use of laser measurement techniques is also found to most suitable for monitoring convection when the growth process is in progress.

2.4. Reconstruction of Three Dimensional Concentration Field Around a Growing Crystal

The use of tomography for the measurement of three-dimensional concentration field around the growing crystal has found limited application in the area of crystal growth. Gatti *et al.* [53] carried out three-dimensional measurements of concentration field using multidirectional holographic interferometry. The authors proposed a mathematical technique for interferometric data processing. The concentration field was reconstructed from the projection data using a numerical algorithm that was based on the "sinc" method developed earlier by Sweeny and Vest [54]. The projections were collected over a limited range of view angles. Notcovich *et al.* [55] investigated the three-dimensional temperature field around heavy ice (D_2O) growing from super cooled heavy water using interference tomography. Heavy water was used for the experiments since dn/dT, the variation of refractive index of normal water with temperature is small (with a maximum in refractive index at 4^oC). Heavy water has a maximum value of refractive index at 7^oC. In addition the authors also measured the two-dimensional concentration field around sodium chloride (NH_4Cl) crystals growing from a supersaturated solution in water. The authors explained the morphological transition of the crystal faces at high growth velocities. In a recent study [56], the authors of the present article have investigated the three-dimensional concentration field around a KDP crystal growing from its aqueous solution using a laser schlieren technique. Growth experiments were conducted in the diffusion as well as convection-dominated regime. Convolution back projection (CBP) tomography algorithm was employed for the reconstruction of the concentration field in the planes above the growing crystal. The study revealed symmetry of the flow field away from the crystal in the convective regime of the growth process. However, the concentration distribution was seen to be unsymmetrical during transition from the diffusion regime towards the establishment of a stable convective plume. This study is described in detail in Section 9.3.

Despite the availability of considerably large amount of literature on the application of optical techniques to investigate the crystal growth processes, the nature of convection and its relationship to solute distribution, crystal growth rates and quality has not been fully understood. The effects of process parameters such as level of supersaturation of the solution, cooling rate, and rotation of the growing crystal on quality and growth rates has not been studied. Optical techniques have predominantly served as the visualization tools to study the growth kinetics and surface morphologies of the crystals growing from their aqueous solutions. The application of tomography based on optical imaging of a variable refractive index field to investigate the three-dimensional nature of convection currents and its effects on the solute distribution around the growing crystal is yet to be fully explored.

3. Objectives of the Present Work

The importance of convection in determining crystal growth rates and crystal quality has been recognized by a variety of authors. Over the past decade, laser measurement techniques have become popular, though primarily as a flow visualization tool. Recent papers, however, have emphasized the possibility of quantitative measurements using optical methods. Against this background, results are discussed in the present article in the following sequence:

1. **Convection in a rectangular cavity.** The literature on refractive index based methods shows that of the three methods, interferometry has been predominantly applied for qualitative as well as quantitative analysis of the thermal fields in buoyancy-driven convection. Schlieren finds potential applications towards qualitative visualization of the flow field and to a limited extent, it has been applied for quantitative studies. Shadowgraph has been extensively used for qualitative imaging. The three techniques, however, have not been jointly compared against a benchmark experimental configuration. The present work compares interferometry, schlieren and shadowgraph techniques for the measurement of temperature distribution in buoyancy-driven convection in a rectangular cavity. The top and bottom walls of the cavity are maintained at uniform temperatures at all times in an unstably stratified configuration. Fluid media considered are air and water. Temperature differences of 5-50 K for air and 3-10 K for water have been employed in the experiments. Over the range of temperature differences considered, flow was seen to become progressively unsteady, finally becoming turbulent.
2. **Comparison of optical techniques for a crystal growing from its aqueous solution.** Interferometry, schlieren and shadowgraph are employed to visualize the convection field around a growing KDP crystal from its aqueous solution. Experiments have been conducted under practically identical conditions. The goal of the study is to explore the suitability of these measurement techniques to image, analyze and interpret the convective field around a growing crystal.
3. **Schlieren study of convection around a crystal driven by buoyancy and rotation.** The specific goals here are to understand (a) changes in convection pattern at various stages of the crystal growth process, (b) role of convection in creating regions of high/low concentration gradients, (c) possibility of controlling the size of these regions by providing crystal rotation, and (d) to demonstrate the suitability of the schlieren technique for quantitative mapping of solutal concentration around the growing crystal. The process parameters studied are cooling rate of the solution and rotation rate of the growing crystal. Crystal rotation is viewed in this study as a means of diminishing the impact of buoyancy. The crystal size plays an influential role in determining the relative importance of buoyancy and rotation. An independent

study of crystal size has also been presented. Schlieren images have been analyzed to correlate the strength of convection currents with the concentration field and its gradients. The crystal quality has been gauged by examining the transparency of the crystal at the end of the experiment.

4. **Three dimensional concentration contours around a growing crystal.** Laser schlieren technique is employed to investigate the three-dimensional concentration field around a KDP crystal growing from its aqueous solution. The convective field is set-up in the growth chamber by inserting a KDP seed into its supersaturated solution followed by slow cooling of the solution. The projection data in the form of two-dimensional schlieren images have been recorded from four different view angles (0, 45, 90, and 135°) by turning the crystal growth chamber. Since a circular growth chamber is employed in the present experiments, the entire width of the growth chamber for a given view angle could not be covered. Hence a single view provides only partial data, corresponding to the area localized in the vicinity of the growing crystal. The integrated values of concentration are obtained by analyzing the light intensity data of the schlieren images. Subsequently, the concentration fields at various horizontal planes above the crystal are reconstructed using the convolution back projection (CBP) algorithm. Owing to limitations in the optical system, the projection data required for tomographic inversion is often incomplete. Specifically, the number of view angles is small, and in addition, the entire field of view may not be scanned. The applicability of the inversion algorithms can then be ensured only when additional tests are conducted to validate the result obtained. Validation with numerically simulated data is reported in this work. The focus of the present study is to examine the symmetry of the concentration field in the vicinity of the growing crystal in the initial stages (diffusion-dominated growth) and in the stable growth regime (recognized by the presence of a steady convective plume rising from the crystal surface). Of interest is the relationship between the morphology of the crystal with the solutal concentration field around it, during the growth process.

The text is organized as follows. Sections 4.1-4.3 describe the apparatus employed in experiments, while Section 4.4 explains the experimental procedure. Sections 5.1-5.4 discuss interferometry, schlieren and shadowgraph individually in terms of the optical layout, principle of operation and data analysis. Section 6 presents a discussion on measurement errors and uncertainty. Sections 7.1-7.3 present principles of tomography, the CBP algorithm, extrapolation techniques for partial data and a validation study against a numerically simulated convection pattern. Section 8 provides detailed results of convection in a rectangular cavity using the three refractive-index techniques. Section 9 shows in greater detail the convection patterns, interaction of buoyancy and rotation, relationship between growth rates and concentration gradients, and preliminary results obtained from tomographic inversion of the schlieren data.

4. Experimental Details

The apparatus used for experiments of the present study are described in the following sections. In experimental procedure followed for growth of KDP crystals is also discussed.

4.1. Convection in a Rectangular Cavity

Experiments are reported for studying fluid convection in a rectangular cavity. The temperature difference maintained across the horizontal surfaces of the cavity leads to unstable density gradients in the fluid medium. In turn, fluid motion in the form of closed loops is initiated in the cavity. The configuration studied, called Rayleigh-Benard convection is a problem of fundamental as well as practical importance. The flow patterns associated with Rayleigh-Benard convection show a sequence of transitions, leading from steady laminar to unsteady turbulent flow. The strength of convection is quantified by the Rayleigh number, a dimensionless parameter. [For a definition, see Nomenclature.] The Rayleigh number is a measure of the ratio of energy released in the buoyancy field to that lost by viscous dissipation. The values realized in the present experiments are quite large in comparison to the critical Rayleigh number of 1707 reported for a horizontal differentially heated infinite fluid layer at the onset of convection (Gebhart *et al.* [2]).

The apparatus used to study the buoyancy-driven convection in a horizontal fluid layer is shown schematically in Figure 2. The cavity is 447 mm long with a square cross-section of edge 32 mm. The test cell consists of three sections namely the top plate, the fluid layer enclosed in a cavity and the bottom plate. The top and bottom walls of the cavity were made of 3 mm thick aluminum plates. The flatness of these plates was manufacturer-specified to be within ±0.1 mm and was further improved during the fabrication of the apparatus. The central portion of the experimental apparatus is the test section containing the fluid medium. The side walls of the cavity were made of a 10 mm Plexiglas sheet. In turn the Plexiglas sheet was tightly wrapped with thick bakelite padding in order to insulate the test section with respect to the atmosphere. The height of the test section was 32 mm and was measured to be uniform to within ±0.1 mm. A window was provided in the direction of propagation of the laser beam. It was held parallel to the longest dimension of the cavity for recording the projected convective field in the form of two-dimensional images. The apparatus was enclosed in a larger chamber made of thermocole to minimize the influence of external temperature variations. The room temperature during the experiments was a constant to better than ±0.5°C over a 10-12 hour period. Experiments were initiated by flowing water over the hot and cold surfaces respectively. After the start of the experiments, the surfaces reached a steady state within a few minutes. The thermal fields in air evolved over a longer period of time and stabilized over 5-6 hours. In water, dynamically steady patterns were realized in 1-2 hours. The thermally active surfaces were maintained at uniform temperatures by circulating a large volume of water over them from constant temperature baths. Temperature control of the baths was rated as ±0.01°C; at the cavity location, direct measurements with a multi-channel temperature recorder showed a spatial variation of less than ±0.1°C. For the upper plate, a tank-like construction enabled extended contact between the flowing water and the aluminum surface. Special arrangements were required to maintain good contact between water and the lower surface of the plate. Aluminum baffles introduced a tortuous path to flow, thus increasing the effective interfacial contact area.

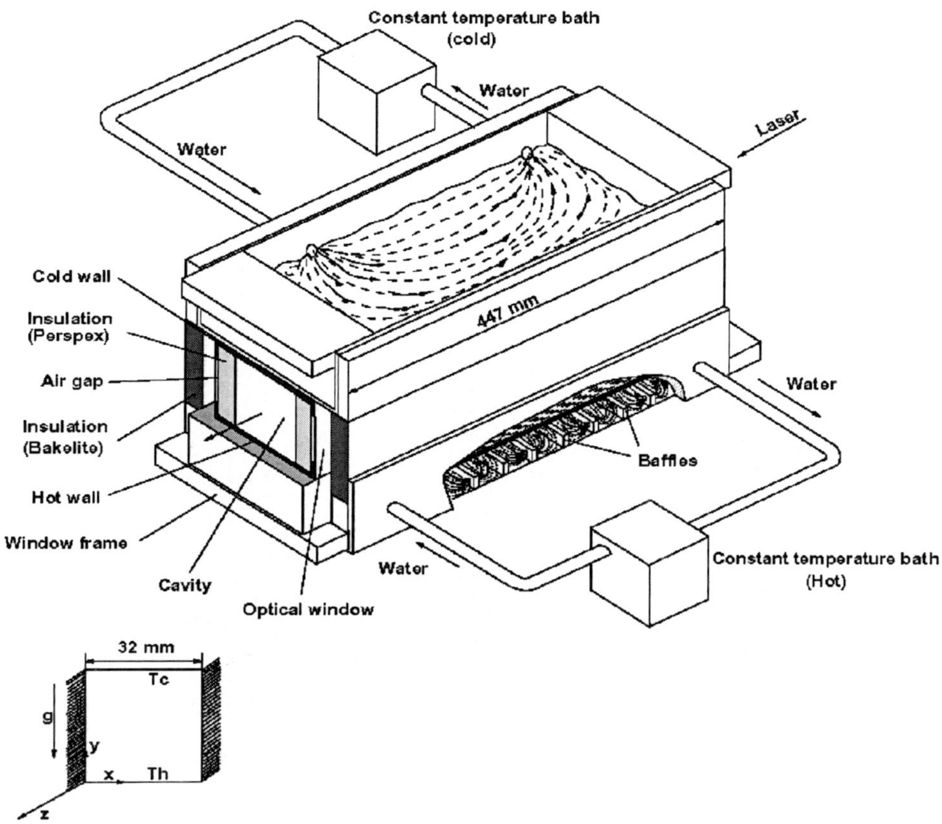

Figure 2. Schematic drawing of the rectangular cavity experimental set up

For air, the temperature differences applied across the cavity walls were in the range of 5-50 K. These correspond to Rayleigh numbers of 14×10^3 to 14×10^4. In experiments with water, the temperature differences applied were in the range of 3-13 K. These correspond to Rayleigh numbers of 25×10^5 to 13.5×10^6 respectively. The Rayleigh numbers realized in the experiments can be seen to be quite large in comparison to the critical value of 1707, indicating the presence of strong fluid motion in the cavity.

4.2. Convection Around a Growing Crystal

The apparatus in which experiments have been conducted on convection around a crystal in a top-hanging arrangement is discussed. The apparatus in which the crystal is held over a platform is described in Section 4.3.

Experiments that compare images of convection around the crystal recorded by interferometry, schlieren, and shadowgraph have been conducted in a growth chamber made of glass (17 cm diameter and 22 cm height) as shown in Figure 3. Glass used for the experimental chamber has the advantage of being a smooth surface on which unwanted nucleation cannot take place. In addition, glass is chemically non-reactive towards the KDP solution. The crystal growth chamber is in turn, placed in a Plexiglas tank. Temperature of the solution inside the growth chamber is maintained at prescribed levels by thermostatically controlled water circulating in the outer tank. The temperature of water is stabilized in the outer tank to within $\pm$ 0.1°C with the help of a programmable temperature controller (*Eurotherm*). A K-type thermocouple wire fixed to the outer surface of the growth chamber provides the feedback to the controller. With this arrangement, ramp rates could be applied to cool the solution over long periods of time. For example, a linear change in temperature of the solution from 25 to 36°C over a period of 60 hours was closely realized in the present set of experiments. Sufficient insulation is provided on the outer surface of the entire assembly to reduce thermal interaction between the growth cell and the ambient.

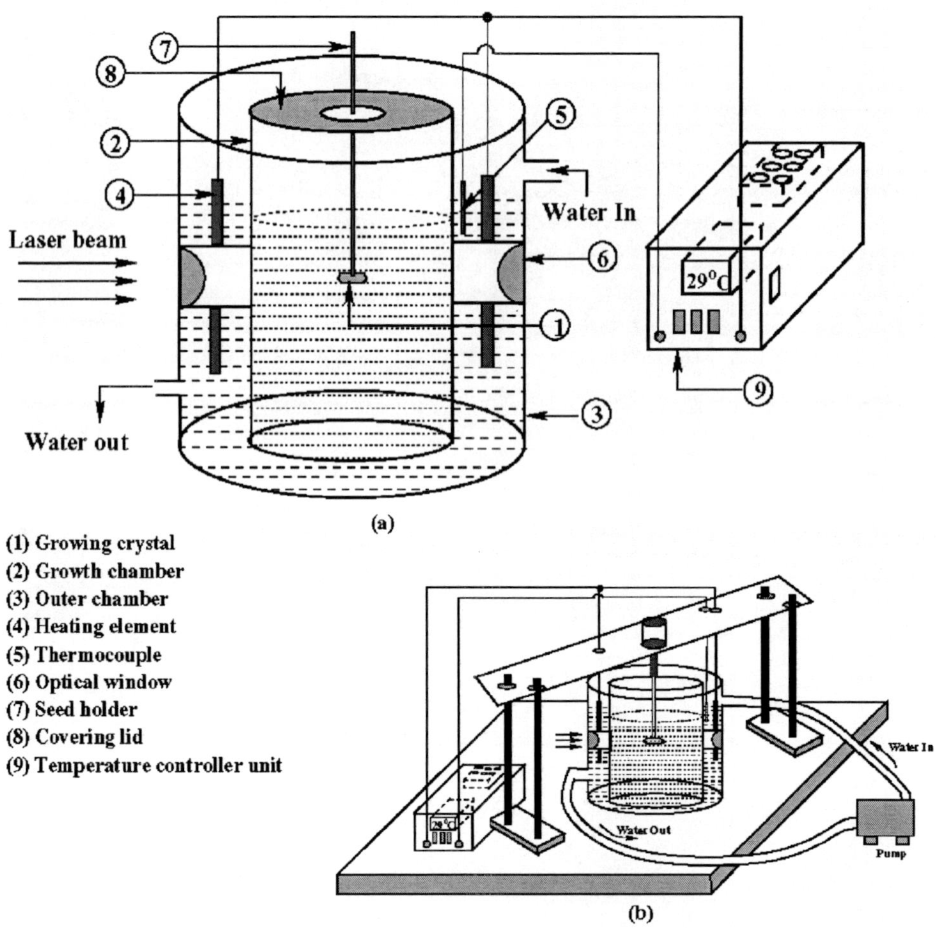

Figure 3. (a) Schematic diagram of the crystal growth chamber, and (b) complete assembly

For visualization of the concentration field by the schlieren technique, circular optical windows (*Stranda*, BK-7, 60 mm diameter, 5 mm thickness, $\lambda/4$ flatness) have been fixed on the walls of the growth chamber at opposite ends. The growth process is initiated by introducing into the growth cell a spontaneously crystallized seed crystal fixed at one end of a thin glass rod. The glass rod (diameter 3 mm) is small enough not to disturb the convection plume rising above the crystal. Rotation to the growing crystal is imparted with the help of a stepper motor (12V, 0.6A, *RS components)*.

For experiments employing interferometry, a double-walled compensation chamber is placed in the path of the reference beam. In order to ensure identical conditions in terms of the undisturbed optical path length, the volumetric capacities of the compensation chamber and growth chamber are kept equal. The inner cell of the compensation chamber contains the KDP solution that has a salt concentration corresponding to saturation conditions at the ambient temperature of 25°C. In the growth process, the temperature range utilized is 25°C-36°C, and no deposition was seen in the experiments anywhere in the cell. The solution in the test section and the compensation chamber are at practically identical temperatures through out the duration of the experiment. This is accomplished by splitting the outlet of the water-circulating pump into two parts, one leading to the growth chamber and the other to the compensation chamber. In the absence of a crystal, the optical path lengths in the test and the compensation chamber could be balanced during the cooling of the solution. Consequently, there was no fringe formation (in the infinite fringe setting mode) and fringe deformation (in the wedge fringe setting mode) from either thermal disturbances or those related to small differences in concentration. Fringe formation in the presence of the crystal occurred due to salt concentration differences in the solution with respect to the average value in the compensation chamber.

In the present set of experiments, the growth of the seed crystal is initiated by slow cooling of the supersaturated solution. Two values of ramp rates have been employed: 0.05°C/hour and 0.1°C/hour. The rate at which the solution is cooled plays an important role in controlling the strength of convection currents, and hence the growth rates and crystal quality. Low rates keep the degree of supersaturation to a small value. The growth process is then diffusion-dominated and is inefficient in terms of the size of the crystal grown. On the other hand, very high values of the ramp rate give rise to vigorous convection currents resulting into deterioration of the crystal quality. At intermediate cooling rates, high crystal quality is to be expected.

Stirring the solution reduces the natural convection-induced temperature oscillations by homogenizing the bulk solution. Hence the importance of optimum rates of rotation in crystal growth processes has gained recognition. The two most widely used stirring mechanisms are the rotation of the seed and/or the rotation of the crucible. In the present work, experiments have been conducted by continuously rotating the growing crystal at 15 rpm.

4.3. Crystal Growth Chamber for Tomographic Projections

Tomographic projection data are recorded by viewing the convection pattern at various angles over the range of 0 to 180°. If the pattern is strictly steady, projections can be recorded by turning the crystal through small incremental angles. This approach has the drawback of creating unsteadiness in the flow field. Results obtained by turning the crystal have been discussed in the context of validating the tomographic algorithm with measured data. An appropriate technique is to keep the growth chamber fixed and turn the axis of the source-detector system. This route could not be followed in the present study owing to the physical size of the laser, CCD camera and the accompanying optical components. Instead, the growth chamber along with the crystal was turned through definite angles to image the concentration field. The curvature of the beaker led to ray-bending problems and prevented scanning of its entire width. The portion of the solution visible through optical windows alone has been used for measurement. Unlike the top-hanging arrangement of Section 4.2 (where the supporting rod disturbed the convective plumes), experiments have been conducted here using a crystal mounted above a platform.

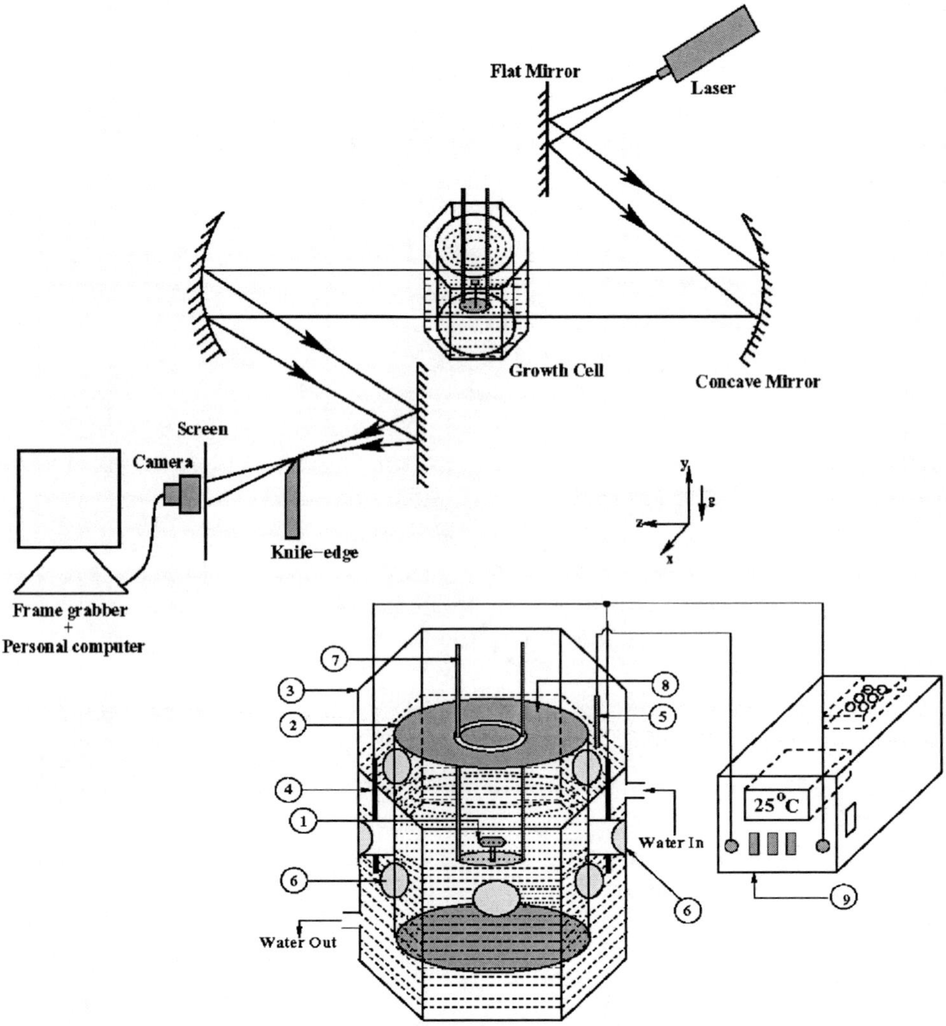

Figure 4. Schematic diagram of the four-view crystal growth chamber placed in the path of the laser beam in a *Z*-type schlieren set up. (1) Growing crystal on a platform, (2) Growth chamber, (3) Outer chamber, (4) Heating element, (5) Thermocouple, (6) Optical window, (7) Seed holder (platform configuration), (8) Covering lid, (9) Temperature controller unit.

The crystal growth chamber used in the present experiment is shown schematically in Figure 4. It comprises of a glass chamber that holds the KDP solution, has a diameter of 16.5 cm with a height of 23 cm. For visualization of the concentration field by the schlieren technique, circular optical windows (BK-7, 40 mm diameter, 5 mm thickness, $\lambda/4$) are fixed on the glass beaker at opposite ends. A total of eight such windows permit passage of the laser beam for four view angles. Parallelism and straightness of the optical windows are crucial for generating meaningful images, and considerable precautions have been taken in this regard. The Plexiglas tank surrounding the growth chamber is octagonal in plan. It ensures large enough volume for the circulating water to keep the KDP solution at the required temperature level over a considerable period of time. Four heating elements placed diametrically opposite in the outer chamber maintain the temperature of circulating water, and hence the KDP solution. Electrical input to the heating elements is regulated by a programmable temperature controller (*Eurotherm*). As in the apparatus discussed in Section 4.2, a K-type thermocouple wire fixed to the outer surface of the growth chamber provides the feedback to the controller. Uniformity of temperature within the solution is ascertained by recording temperatures at various locations using 26 gage K-type thermocouples. With this arrangement, it was possible to reduce the temperature of the solution from 36 to 25°C linearly with time over a time frame of 60 hours.

The salt-solution has an initial temperature that is high enough to keep it from becoming fully saturated. A KDP seed crystal spontaneously crystallized in a second vessel is placed on a glass platform and introduced in the growth cell. When the bulk temperature is lowered, the solution becomes super-saturated with salt that in turn deposits on the crystal. The solution adjacent to the crystal is now close to being saturated with the dissolved salt. A crystal growing from its aqueous solution thus creates a three-dimensional solute distribution in its vicinity. The solutal concentration gradients and the accompanying gradients in density of the solution are responsible for the appearance and evolution of buoyancy-driven convection currents in the growth chamber.

4.4. Experimental Procedure

The crystal growth experiments were performed in the following manner. A supersaturated solution of KDP powder is prepared in distilled, de-ionized water. The amount of KDP salt to be dissolved is determined from its solubility curve [57] at an average temperature of 35°C. The solution is stirred long enough to ensure complete dissolution of the solute. It is filtered using Whatman-100 filter paper to remove residual microscopic particles. The solution is maintained at relatively high temperature for about 4-5 hours using water circulation from the temperature control unit. Subsequently, the solution is cooled by applying a suitable ramp rate, 0.1°C/hour in the present study. When the solution reaches a value right enough for it to become saturated, a KDP seed is introduced into the solution using a thin glass rod. The orientation of the seed crystal is adjusted in such a way that the prismatic faces ({100}) are placed normal to the direction of the laser beam. Experiments are allowed to run for a considerably longer duration of time (≈60 hours) in order to allow the growth of large crystals. Precautions are taken to avoid external disturbances and vibrations that may introduce unsteadiness in the flow field. For the experiments discussed below, the crystal size increased from 0.3 cm^3 to 3 cm^3 in 60 hours, while preserving high crystal transparency.

5. Laser Measurement Techniques and Data Analysis

The optical configurations for interferometry, schlieren and shadowgraph as well as data retrieval from optical images are discussed in the following sections (see Goldstein [58] and Settles [59] for a longer discussion.).

5.1. Optical Arrangement

For measurements, a continuous wave helium-neon laser (Spectra Physics, 35 mW) has been used as the coherent light source. A monochrome CCD camera (*Sony*, Model: XC-ST50) of spatial resolution of 768×574 pixels was used to record the convective field. The camera was interfaced with a personal computer (HCL, 256 MB RAM, 866 MHz) through an 8-bit A/D card. Interferograms, schlieren and shadowgraph images presented in this work have been subjected to image processing operations to improve contrast; the data analysis for recovering the field concentration is, however, based on the intensity changes in the original images.

The layout for each of the three imaging techniques used in the present work is shown in Figure 5. The Mach-Zehnder interferometer used in the present study is shown in Figure 5(a), and has been described earlier by Mishra *et al.* [19,60]. It has two mirrors and two 50% beam splitters of 150 mm diameter. The mirrors are coated with 99.9% silver and employ a silicon dioxide layer as a protective layer. The interferometer floats on pneumatic legs to isolate the optics from external vibrations. Experiments have been carried out in the infinite as well as the wedge fringe setting [58]. In the infinite fringe setting, the optical path difference between the test and the reference beams is initially zero. When a density disturbance is introduced in the path of the test beam, it is seen as a set of fringes over which the depth-averaged concentration field is a constant. In the wedge fringe setting, the optics is slightly misaligned to produce a set of straight fringes. When exposed to a thermal or concentration field, the fringes are displaced to an extent depending on the change in temperature or concentration. The fringes in the wedge fringe setting of the interferometer are thus representative of the temperature variation in the rectangular cavity and concentration profile in the crystal growth chamber. In the present study, the infinite fringe setting initially produced an image of uniform brightness indicating true constructive interference between the test and the reference beams. In the wedge fringe setting, the fringes were seen to be straight and horizontal.

The schlieren system used in the present work is of the ***Z***-type, as shown in Figure 5(b). The optics includes concave mirrors of 1.30 m focal length and 200 mm diameter. Relatively large focal lengths make the schlieren technique sensitive to the concentration gradients [58-59]. The knife-edge is placed at the focal length of the second concave mirror. It is positioned to cut off a part of the light focused on it, so that in the absence of any optical disturbance, the illumination on the screen is uniformly reduced. The initial intensity values in the experiment were chosen to be less than 20, on a gray scale of 0-255. The knife-edge is set perpendicular to the direction in which the density gradients are to be recorded. In the present study, the gradients are expected to be predominantly in the vertical direction of gravity, and the knife-edge has been kept horizontal.

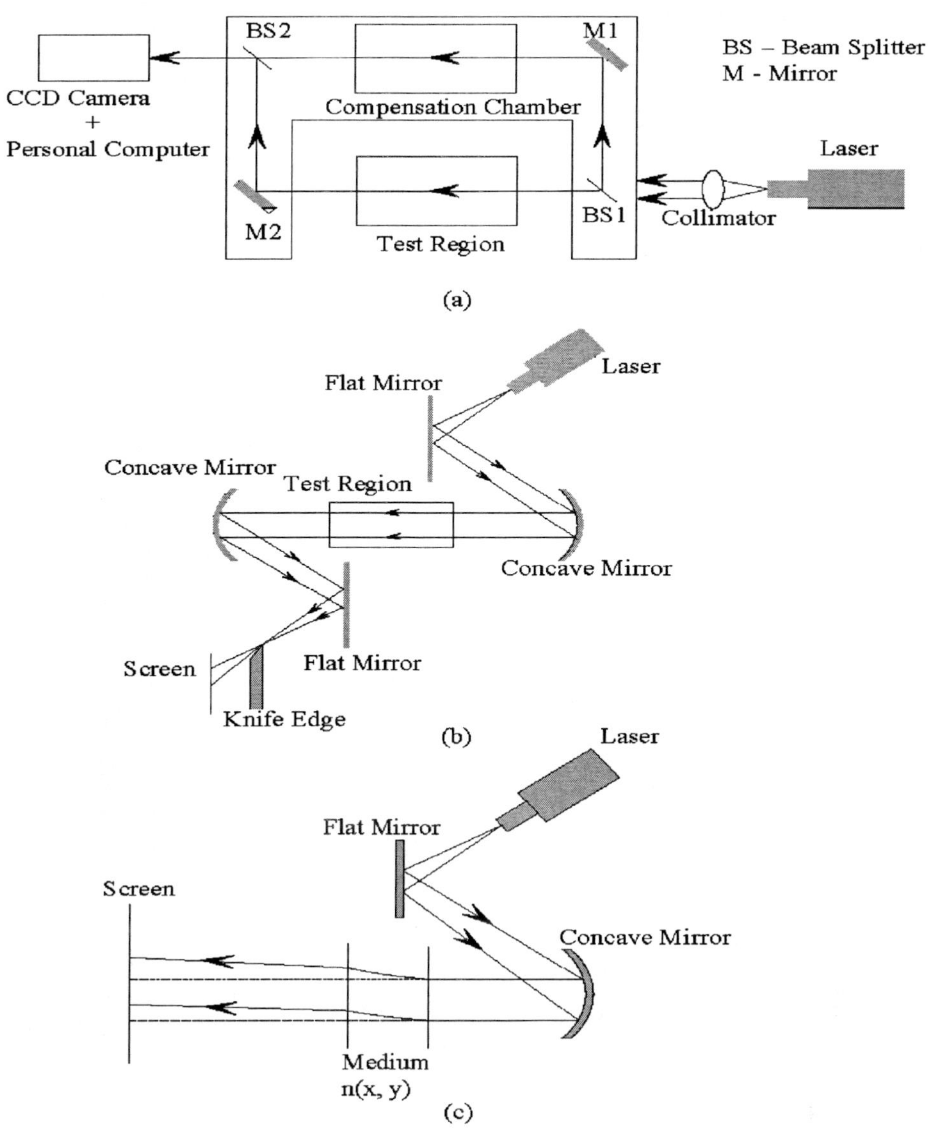

Figure 5. Optical layout of a (a) Mach-Zehnder interferometer, (b) Schlieren and (c) Shadowgraph.

Shadowgraph images have been recorded using the same optical components as employed in the interferometry setup by first blocking the reference beam and allowing the test beam alone to fall on the screen (Figure 5(c)). Shadowgraph images have also been recorded from the schlieren apparatus by collecting the refracted light beam emerging from the test cell. The position of the screen on which the shadowgraph images are displayed plays an important role in data analysis [9, 58]. The screen position is chosen so as to improve the image contrast, while extracting the dominant features of the flow field. Beam refraction from regions of high concentration gradients can interfere with those passing through one of nearly constant concentration. This factor is taken into account while fixing the camera position. The initial intensity distribution in shadowgraph experiments before inserting the seed corresponds to the Gaussian intensity variation of the laser source itself.

5.2. Image Processing

The three optical methods generate images that are path integrals of the refractive index fields and in turn, the concentration (or temperature) distribution in the fluid medium. The integrals can be simplified if the fields being studied are strictly two-dimensional. In general, the local information can be extracted using principles of tomography [61-62]. In the present work, the images have been interpreted as carrying information of the solutal concentration (or thermal) field that is an average along the direction of propagation of light.

The determination of the concentration/temperature field from interferograms requires several intermediate steps, including (a) noise removal, (b) edge detection, (c) location of intensity minima within fringe bands, and (d) fringe thinning. Step (d) involves fitting a smooth function through points of intensity minima within a single fringe, as obtained in step (c). In addition, assigning temperatures to fringes, followed by transferring the data to a Cartesian grid are important. The fact that fringe thickness is small in regions of high concentration or temperature gradients has not been used in the present work. The analysis and interpretation of fringe patterns in interferometry has been discussed in detail by various authors [63, 64]. The approach described in [64] has been implemented in the present study.

In contrast to interferometry where information is localized at the fringes, a schlieren image carries information related to the local temperature/concentration in the form of an intensity distribution. The advantage here is that data is available at the pixel-level of the camera. Drawbacks include the errors due to superimposed noise associated with scattering and the possibility of device saturation. In the present work, the first factor is taken to be less significant because the field variable is obtained by integrating the intensity distribution (see Section 5.3), an operation that tends to smooth noisy profiles. The second factor was circumvented by reducing the laser intensity using neutral density filters. The local temperatures (and concentrations) were then determined by numerically integrating the appropriate governing equation in the direction normal to the knife edge.

5.3. Data Reduction

The present section examines the suitability of the three optical techniques of the present study for quantitative analysis of the convective field within a rectangular cavity and during crystal growth. An aspect shared by the three techniques is that they generate projection data, namely information that is integrated in the direction of propagation of the light beam. The result of analysis is thus a concentration (or temperature) that is ray-averaged over the length L of the growth chamber. The steps involved in the analysis of the experimental images are briefly discussed below.

Refractive index techniques depend on the fact that for a transparent material, refractive index and density share a unique relationship, called the Lorentz-Lorentz formula [58]. In addition, if pressure changes within the fluid region are small, density will be uniquely related to solute concentration and temperature. The functionality is often linear since changes in concentration and temperature are often small in many applications. In rectangular cavity experiments, temperature is the only dependent variable and density (and hence refractive index) relates to temperature. Though the crystal growth process is driven by the cooling process of the solution, the ramp rate is often small enough for thermal equilibrium to prevail for short periods of time through out the experiment. In the present analysis, it is assumed that the aqueous solution has negligible temperature gradients. Hence, refractive index becomes a measure of concentration itself. The material property that determines the sensitivity of the optical measurement is dn/dC, where n is the refractive index and C is solute concentration.

The following discussion is presented in terms of solutal concentration. It is equally applicable to imaging thermal convection.

Interferometry: For a light source of wavelength λ, the change in concentration required per fringe shift in the infinite fringe setting is given by the equation [58]

$$\Delta C_{\varepsilon} = \frac{\lambda / L}{dn / dC} \tag{1}$$

The fringe positions are to be determined from interferogram analysis. In the wedge fringe setting, it can be shown that the fringe displacement from the initial position is proportional to the change in concentration with respect to the portion of the solution where the fringes are undisturbed. These results hold under the approximation that refraction effects are small.

Schlieren: Image formation in a schlieren system is due to the deflection of light beam in a variable refractive index field towards regions that have a higher refractive index. In order to recover quantitative information from a schlieren image, one has to determine the cumulative angle of refraction of the light beam emerging from the growth chamber as a function of position in the x-y plane. This plane is defined to be normal to the light beam, whose direction of propagation is along the z-coordinate. Using principles of ray-optics, the total angular deflection δ can be expressed as [58-59]

$$\delta = \frac{1}{n_a} \int_0^L n \frac{\partial(\ln n)}{\partial y} dz$$

where n is the refractive index at any point in the solution. The change in the intensity field (ΔI) relative to the background intensity distribution (I_k) can now be related to the refractive index field directly as

$$\frac{\Delta I}{I_k} = \frac{f}{a_k n_a} \int_0^L \frac{\partial n}{\partial y} dz \tag{2}$$

where n_a (~1) is the refractive index of the ambient, a_k is the size of the focal spot at the knife edge, and f is the focal length of the de-collimating mirror. This equation shows that the schlieren technique records the average gradient of refractive index over the path of the light beam. In terms of the ray-averaged refractive index, the governing equation for the schlieren process can be derived as:

$$\frac{\Delta I}{I_k} = \frac{f}{a_k} \frac{\partial n}{\partial y} L \tag{3}$$

where L is the length of the path traversed by the light beam through the growth chamber, namely its diameter. Equation 2 requires the approximation that changes in the light intensity occur due to beam deflection, rather than its physical displacement. The concentration gradient is related with the refractive-index gradient of the KDP solution using the following relation [30]:

$$\frac{\partial N}{\partial y} = \frac{9n}{2\alpha(n^2+2)^2} \frac{\partial n}{\partial y} \tag{4}$$

Here N is the molar concentration of the solution and α is the polarizability of the KDP crystal (=4.0 cm^3/mole). Combining Equations 3 and 4 and integrating from a location in the bulk of the solution (where the gradients are negligible), the concentration distribution around the growing crystal can be uniquely determined.

The contribution of refraction of light at the confining optical windows has been accounted for by applying a correction factor in Equation 3 for the angle of deflection with which the beam emerges from the growth chamber. It can be shown using Snell's law that the correction factor is equal to the refractive index of the KDP solution (=1.355) at the ambient temperature.

Shadowgraph: The shadowgraph arrangement depends on the change in the light intensity arising from beam displacement from its original path. Shadowgraph analysis requires tracing the path of individual rays through the aqueous solution. When subjected to linear approximations that include small displacement of the light ray, a second order partial differential equation can be derived for the refractive index field with respect to the intensity contrast in the shadowgraph image [57]. With D as the distance of the screen from the optical window on the beaker and Δ as the Laplace operator in the x-y plane, this equation is

$$\frac{\Delta I}{I_k} = LD\left[\Delta\{\ln n(x,y)\}\right] \tag{5}$$

Equations 3 and 5 have to be suitably integrated to determine the refractive index, and hence the concentration field. Integration of the Poisson equation (5) can be performed by a numerical technique, say the method of finite differences. When the approximations involved in Equations 3 and 5 do not apply, optical techniques can be used for flow visualization alone.

In interferometry, one can measure concentration values only at those points where fringes appear. The quantity of information is also limited by the number of fringes (infinite fringe setting) or their deformation (wedge fringe setting). Schlieren and shadowgraph generate information about the refractive index field at every pixel of the image in the form of intensities.

6. Uncertainty and Measurement Errors

Errors in the experimental data are associated with misalignment of the apparatus with respect to the light beam, noise generated at different stages of the experiment, including the imperfections of the optical components, and the intrinsic unsteadiness of the convection process itself. All experiments were conducted several times to establish the repeatability of the convective patterns, for the parameter range of interest. A quantitative assessment of errors and uncertainty was possible in experiments on convection in a rectangular cavity because of knowledge of the lower and upper surface temperatures.

A series of cross-checks have been enforced to validate the quantitative results obtained from experiments in the rectangular cavity. These include predicting the temperature of the cold top wall by starting from the lower hot wall, as well as comparing steady heat transfer rates across the cavity against published correlations. For example, an interferometry experiment with hot and cold temperatures of 305.5 K and 300.5 K yielded the cold wall temperature as 300.2 K. For temperatures of 300 and 320 K, the hot wall temperature was predicted to be 319.65 K. In schlieren, temperatures of 300 and 320 K were predicted as 301 and 319 K. The errors were considerably smaller at lower temperature differences across the cavity where the flow field was clearly steady. The comparison with correlations of average heat fluxes across the cavity for various temperature differences is discussed in Section 8.

In experiments with air, the time-dependent movement of the fringes during interferometry and the time-wise change in intensity for schlieren were not severe enough at the lower end of temperature differences, specifically the lower end of the Rayleigh number range. For $Ra>51{,}000$ unsteadiness was a source of uncertainty in the local temperature profiles as well as the wall heat transfer rates. Quite often, unsteadiness was a source of measurement uncertainty in air. It was less serious in water as well as the aqueous solution of KDP. Analysis of convection patterns in the KDP solution could be carried out under conditions of mild unsteadiness by averaging a time-sequence of images. The results obtained in the present work can be taken to be quantitatively meaningful for the lower range of Rayleigh numbers. For very high Rayleigh numbers, the images are purely representative of the instantaneous thermal/solutal fields in the fluid region.

An additional check carried out in the crystal growth experiments was one of solutal mass balance. Since the convective field was imaged from various angles, it was possible to independently determine the average concentration of KDP on a given plane. These averages were within 0.1% of each other as long as the convection plume was steady. Larger discrepancies were seen in those experiments where the convection process was strongly unsteady. Only those images that satisfied the mass balance check were used for tomographic reconstruction.

7. Three Dimensional Reconstruction from Schlieren Data

The schlieren image yields gradients of refractive index and hence concentration. As discussed in Section 5.3, the gradient information can be integrated to yield data in terms of concentration itself. Subsequently, the concentration field on selected horizontal planes above the growing crystal can be reconstructed using principles of tomography [61-62]. For this purpose, the left side of Equation 2 is interpreted as projection data of the refractive index gradient.

The schlieren images have been utilized to reconstruct the concentration field on horizontal planes above the crystal. For this purpose, images have been recorded at four different view angles. Since these images are time-separated, the present study is restricted to a steady convection conditions. Minor changes in the concentration field with time have been accounted for by averaging a sequence of four successive images. The schlieren images recorded during the initial transients as well as late stages of growth when convection is unsteady have not been utilized for three dimensional reconstructions.

Application of interferometric tomography for buoyancy-driven convection has been discussed earlier by Mishra *et al.* [19]; also see the related review article [64].

7.1. Computed Tomography

Tomography has been used in the present work to reconstruct two-dimensional concentration fields over individual horizontal planes from their one-dimensional projections. The third dimension is filled by moving from one plane to the next in the third (vertical) direction. The recording configuration shown in Figure 6 is the parallel beam geometry. The convolution back projection (CBP) algorithm has been used in the present study for tomographic reconstruction. Significant advantages of this method include its non-iterative character, availability of analytical results on convergence of the solution with respect to the projection data, and established error estimates [61-62,65]. The treatment of reconstruction for partial data adopted in the present work differs from that of Mishra *et al.* [20] where an iterative approach has been followed.

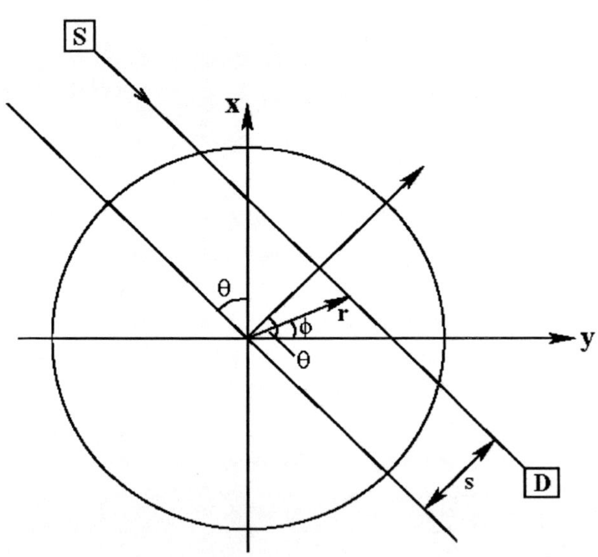

Figure 6. Schematic drawing of data collection using parallel beam geometry. S-source, D-Detector, s-perpendicular distance from the center of the object to the ray, θ-view angle, and (r,φ)-polar coordinates.

In an experiment, projection data can be recorded either by turning the crystal growth chamber or the source-detector combination. The latter is particularly difficult due to stringent requirement of alignment. With the first option, it is not possible to record a large number of projections, owing to the inconvenience of installing plane optical windows in a circular beaker. Further, for a cylindrical growth chamber, the entire field of interest cannot be imaged due to the curvature of the test cell. Instead, the central core region (corresponding to the size of the optical windows), that includes the growing crystal, has been recorded. In this respect, the projection data set is incomplete. In order to generate a complete projection data set for each view angle, an extrapolation procedure has been adopted and is discussed in Section 7.2.

Projection data has been recorded for four view angles of 0, 45, 90, and 135° in the present experiments. The data for 180° is taken to be identical to that for zero angle. Information for intermediate view angles has been generated by employing linear interpolation on the experimentally recorded data.

7.2. Extrapolation Scheme

The experimental set up employed in the present work enables only a part of the aqueous solution in the beaker to be scanned by the laser beam, being limited by the size of the optical windows. The projection data is thus incomplete, as shown schematically in Figure 7. Specifically, the experimental data in the form of the schlieren images covers about 25% of the beaker diameter. In order to successfully apply the CBP algorithm for tomographic inversion, one needs projection data over the entire width of the physical domain for each view angle. In the present work, the experimentally recorded partial data has been extrapolated to derive information about the portion of the solution beyond the optical windows. In all experiments, the region inclusive of the crystal has been imaged; extrapolation is applied to the portion of the field from the edges of the laser beam to the walls of the beaker.

The applicability of extrapolation to the present study can be justified on the basis of two factors: First, the concentration level away from the growing crystal, corresponding to the supersaturated solution is practically constant. This is confirmed in the recorded schlieren images where the changes in the intensity are found to be localized in the vicinity of the growing crystal alone. These images are discussed in the subsequent sections. Secondly, the information content of schlieren images decrease with the geometric path length of the light beam within the beaker, as indicated by the integration limits in Equation 2. The diminishing chord length of the beaker towards the edges shows that the measurement procedure de-emphasizes concentration variation occurring towards the sides of the beaker. Specifically, the chord length is zero at the extremities of the beaker. The approximation involved in extrapolating concentration outside the measurement volume is thus expected to be less serious in predicting the concentration field closer to the center of the beaker and hence the crystal location. Further justification can be based on the fact that concentration is continuously distributed in the fluid volume.

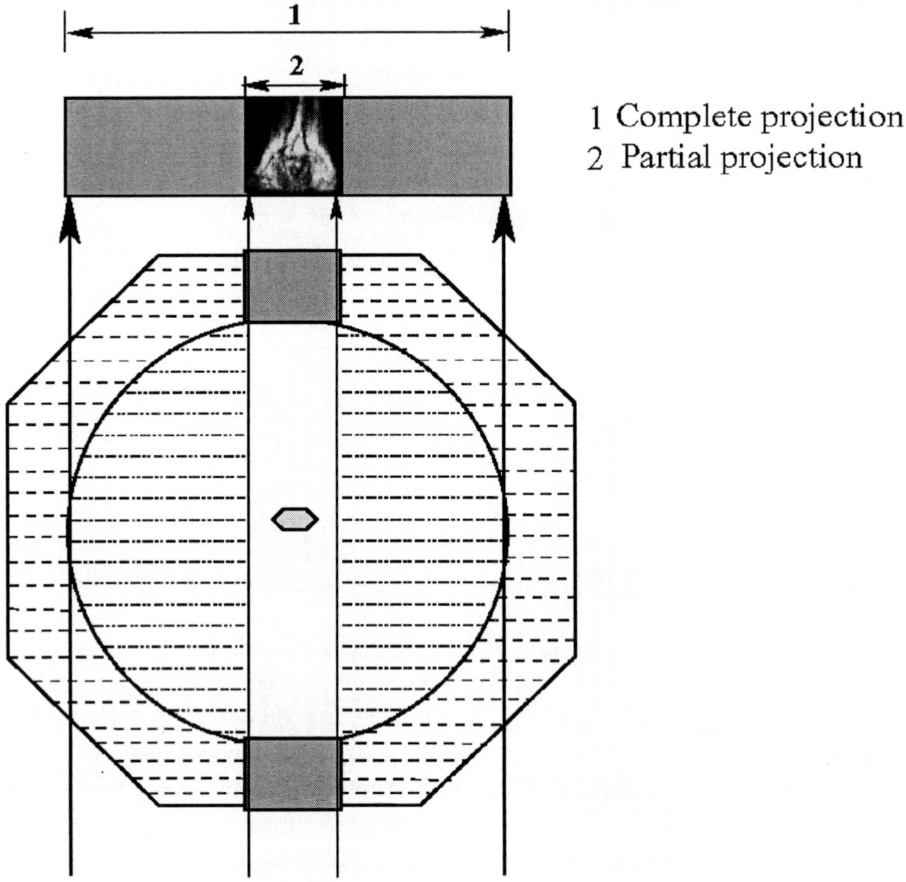

Figure 7. Definition of partial projection data

In the present work, a tenth order polynomial has been used to extrapolate the concentration distribution, starting with the portion covered by the optical windows. Polynomials of order 5 to 10 produced practically identical results. The limiting values of concentration in the far field, and the necessity of maintaining slope-continuity in the concentration distribution at every point have been enforced. An independent check on the accuracy of the experiment, data analysis and extrapolation is the conservation of solutal mass in each of the projections. Mass balance was found to be better than 0.01% in all the experiments analyzed using tomography. Minor imbalances are subsequently rectified by normalizing the data-set. To account for changes in the integration length along the direction of the laser beam, the extrapolated data is multiplied by the local chord length as one proceeds from the center towards the periphery of the growth chamber. This projection data is used for tomographic reconstruction of the concentration field over a horizontal plane of the solution. The above approach has been successfully tested in the context of numerically simulated buoyancy-driven convection and is discussed in Section 7.4.

7.3. Convolution Back-Projection

In the convolution back-projection algorithm (CBP), the reconstructed function, $f(r,\phi)$, is evaluated by the integral formula [61,62,65]:

$$f(r,\phi) = \int_{0}^{\pi} \int_{-D/2}^{D/2} p((s' - s);\theta)q(s)dsd\theta \tag{6}$$

where

$$q(s) = \int_{-Rc}^{Rc} | R | W(R)\exp(i2\pi Rs)dR$$

Here $p(s;\theta)$ is the projection data and s is the perpendicular distance of the data ray from the center of the object. In addition, θ denotes the source-detector line with respect to a fixed axis (and hence the view angle), D is the diameter of the growth chamber, and s' is the s-value of the data ray passing through the point (r,ϕ). The symbol R is the Fourier frequency, $q(s)$ is the convolving function of Equation 6, and $W(R)$ is the filter function. Also see Figure 6 for an explanation on the notation used. The filter function vanishes outside the interval $[-R_c,+R_c]$ and is an even function of R. Here R_c is the Fourier cut-off frequency and is taken to be $1/2\Delta s$, Δs being the ray spacing. The reconstruction obtained is specific to the choice of the filter function [61-62]. A Hamming filter h54 has been used in the present study. For $|R| < Rc$, it is given by the formula

$$W(R) = 0.54 + (1 - 0.54)\cos\left(\frac{\pi R}{R_c}\right)$$

As discussed by Natterer [62] and Munshi [65], this filter emphasizes the smooth features of the concentration variation, while suppressing small-scale (fine structure) fluctuations. This is quite appropriate to the present study for the following reason. Density variations in the solution arise primarily because of the deposition of the solute on the growing faces of the crystal as the solution is cooled at a given rate. Since the crystal growth rate is slow, one encounters density variations that are distributed in the entire solution, while rapid fluctuations in concentration do not appear [44-45]. Hence, it is of interest to reconstruct the dominant pattern in the concentration field rather than its secondary features.

7.3. Validation of Reconstruction Procedure with Simulated Data

The goal of the present work is to obtain concentration distribution on selected planes above the crystal growing from its aqueous solution from the schlieren images. These images require to be extrapolated to fill the width of the beaker. The extrapolation step combined with the convolution back-projection algorithm is first validated against simulated data. The physical problem considered is buoyancy-driven convection in a differentially heated circular fluid layer with upper and lower walls maintained at specified temperatures. The side wall that is circular is thermally insulated. The fluid considered is air and the Rayleigh number based on the height of the fluid layer is set at Ra=12,000. The temperature distribution in the fluid layer has been obtained by numerically solving the governing equations of flow and energy transport on a fine grid. For definiteness, the thermal field is taken to be axisymmetric; accordingly the isotherms on individual planes of the fluid layer are circular. The physical realizability of axisymmetric convection in a circular cavity around a Rayleigh number of 12000 has been discussed at length by Velarde *et al.* [66] and Leong [67].

With the solution for temperature determined numerically, the projection of the thermal field is obtained by path integration. The accuracy of reconstruction with partial data has been examined in the present study against the available numerical solution.

Errors have been reported in the present section on three grids, namely 64×64, 128×128 and 256×256. Here the first number represents the number of view angles along which projections have been recorded, and the second indicates the number of rays for each view. The definitions of errors considered are:

$$E_1 = \max[(T_{orig} - T_{recon})] \quad \text{absolute maximum temperature difference}$$

$$E_2 = \sqrt{\frac{1}{N}\sum\left[T_{orig} - T_{recon}\right]^2} \qquad \text{RMS error}$$

Here T_{orig} and T_{recon} are the original and reconstructed temperature fields respectively and N is the total number of grid points on the reconstructed plane. All temperatures generated by simulation are dimensionless and fall in the range of 0 to unity. The difference between error norms E_1 and E_2 arises from the fact that the former highlights large isolated errors, while the latter reveals trends that are applicable for the entire cross-section.

The validation of the reconstruction procedure with simulated data is summarized in Figure 8. Figure 8(a) shows isotherms of the thermal field in the fluid layer for a given view angle. This data is presented in the form of contours of the path integrated temperature field. Since the thermal field is axisymmetric, the projection data for all other view angles are identical to Figure 8(a). The reconstruction over a horizontal plane of the fluid layer is shown in Figure 8(b). Results obtained with complete projection data (marked 100% in 8b(i)) and partial data (60% in 8b(ii) and 30% in 8b(iii), symmetrically placed about the center) are also shown. The axisymmetric nature of temperature distribution is brought out in all the reconstructions. This can be taken as a validation of the extrapolation procedure used to convert partial to an approximate but complete data set. A quantitative comparison of the reconstructed temperature profiles along the diameter of the cavity for different combinations of rays and views are shown in Figure 8(c). Profiles obtained with full as well as partial data are reported. For the complete data set, a perfect match between the original and reconstructed profiles can be seen for grid sizes of 128×128 and 256×256, while small errors are to be seen for the 64×64 grid. The extent of deviation from the original increases as the fraction of incomplete data increases. Noticeable errors are to be seen when only 30% of the original data is used, the rest of it being derived by extrapolation. Errors in reconstruction were found to be significantly higher when the partial data set was used without extrapolation.

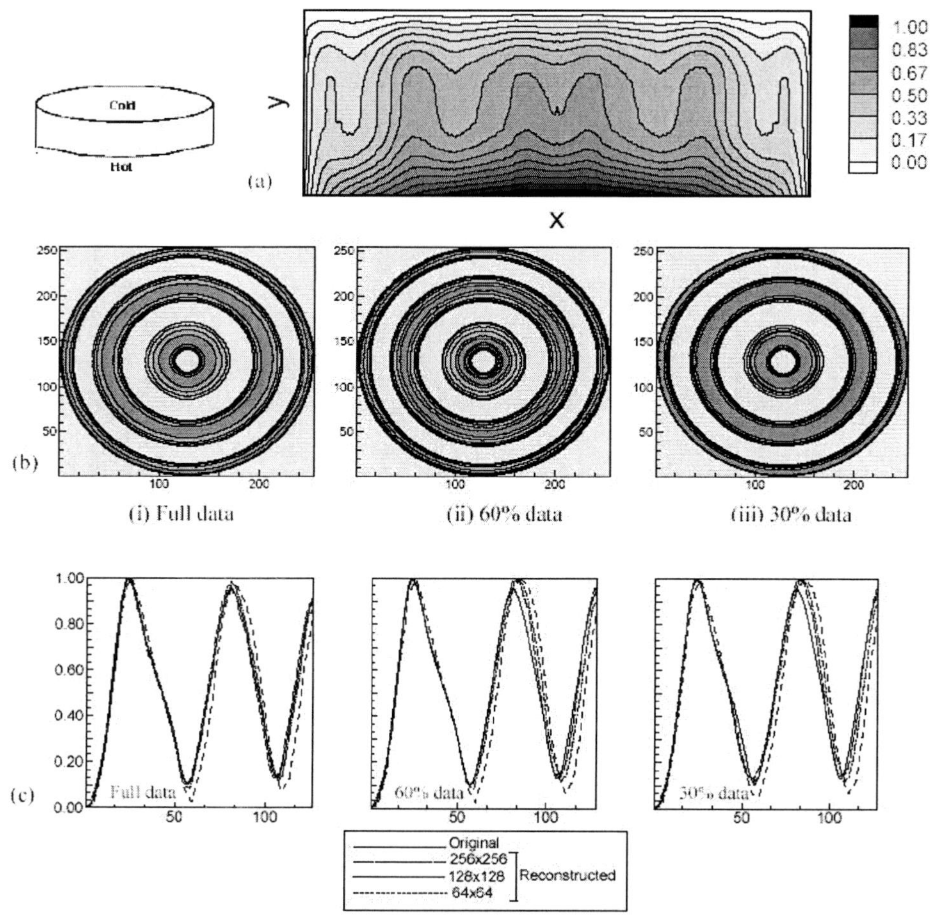

Figure 8. Buoyancy-driven convection in a differentially heated circular cavity. (a) Complete projection data in the form of isotherms for the differentially heated circular fluid layer; (b) Reconstructed temperature contours at y/H=0.65 for full (i) 100% and partial (60% (ii), 30% (iii)) projection data; and (c) Comparison of original and reconstructed non-dimensional temperature distribution along the radial direction for the three different combinations of rays and views.

The magnitudes of errors as a function of discretization of the fluid layer and size of the partial data set are summarized in Table 1. Since the difference between the minimum and maximum temperatures is unity, the percentage error is obtained as $100 \times E_1$ and $100 \times E_2$. In Table 1, error E_1 is consistently seen to be higher than E_2, the latter being an average over the entire field. Both errors decrease as the grid size (number of rays and views) increases. For a given grid size, errors increase as the fraction of original data used in reconstruction decreases. When only 30% of the original is used (the rest being obtained by extrapolation), the maximum errors on a 64×64 grid are 17.3% (absolute maximum) and 7.6% (RMS).

Figure 8 shows that the corresponding reconstructions are qualitatively meaningful, and hence these error magnitudes may be taken to be within limits.

Table 1. Comparison of the original and reconstructed temperature fields in terms of errors E_1 and E_2 for buoyancy-driven convection in a circular cavity

Date Type		Rays×Views	E_1	E_2
Full Data		256×256	0.052	0.028
		128×128	0.109	0.056
		64×64	0.124	0.058
Partial Data	60%	256×256	0.095	0.039
		128×128	0.148	0.061
		64×64	0.152	0.067
	30%	256×256	0.122	0.051
		128×128	0.148	0.072
		64×64	0.173	0.076

8. Benchmark Experiment: A Comparison of Interferometry, Schlieren and Shadowgraph

A direct comparison of images of convection seen in interferometry, schlieren and shadowgraph is presented here. The physical experiment considered is a differentially heated fluid layer enclosed in a rectangular cavity (Section 4.1). In the initial discussion, the temperature difference across the cavity is 10 K in the experiments, while the fluid medium in the cavity is air. The corresponding Rayleigh number has been calculated to be 6×10^4. The interferograms, schlieren and shadowgraph images are compared on the right column of Figure 9. The fringes of an interferogram are lines of constant temperature. With reference to the discussion in Section 5.3, temperature here is to be interpreted as an integrated value over the length of the cavity. It should be noted that fringe shapes in a circular geometry are harder to interpret owing to a change in the geometric path length in a direction the path of the light beam [68-69]. The spatial coordinates of the fringes can be used to obtain the temperature profile. For schlieren and shadowgraph, the information regarding the thermal field is contained in the light intensity variation. The respective thermal properties recovered are the local values of the first derivative and the Laplace operator applied to the temperature. These quantities have been plotted for the mid-plane of the cavity on the left side of Figure 9. The individual data points are specific to the mid-plane of the cavity, while the solid line indicates the overall trend. The shaded circles of the left column indicate the gradients calculated from interferometric data (for schlieren) and from schlieren data (for shadowgraph). A good overall match is a confirmation of the result that schlieren is a derivative of the interferometric field, and the shadowgraph in turn is the derivative of the schlieren. The appearance of dense fringes near the horizontal walls is indicative of high temperature gradients at these locations. This is brought in the schlieren image in the form of an increase in intensity as well as the data points. The central region is a zone of nearly constant temperature, where the gradients (and the light intensity values) are close to zero. Thus, the schlieren images and interferograms correlate quite well with each other. They also correlate with the shadowgraph, once it is realized that in this approach, light is redistributed over the image. In a shadowgraph image, light from the region close to the cold top wall deflects towards the lower hot wall, where the light intensity shows a maximum. Thus, high gradients are represented in a shadowgraph by regions of very low as well as very high light intensity. In the central core, the change in light intensity with respect to the initial setting is small. Thus the Laplacian operation of temperature in this region yields a practically zero value. The thermal lensing effect that distorts the shape of the cavity cross-section is most visible in the shadowgraph.

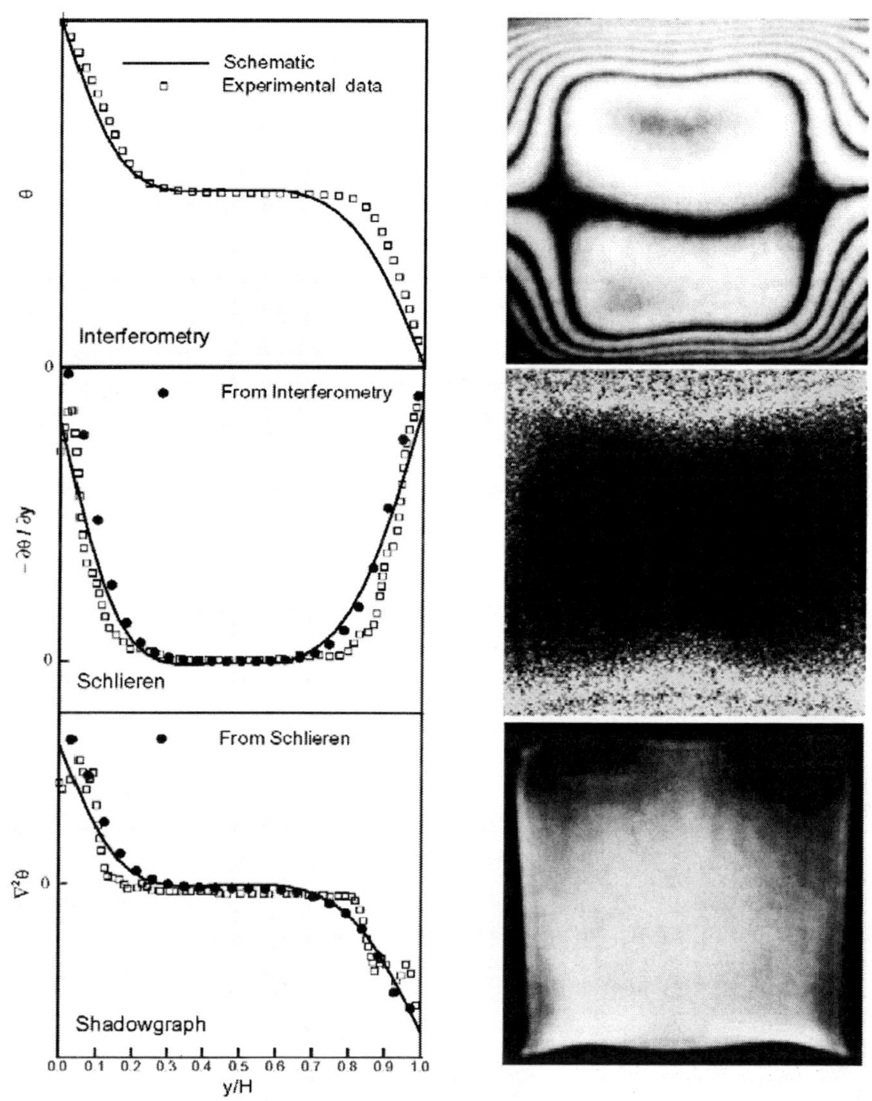

Figure 9. Comparison of the data recovered from the three optical techniques (left column). The corresponding experimental images are shown in the right column. Ra=6×10^4. The solid line in the left column is representative of the trend seen at any vertical section of the image.

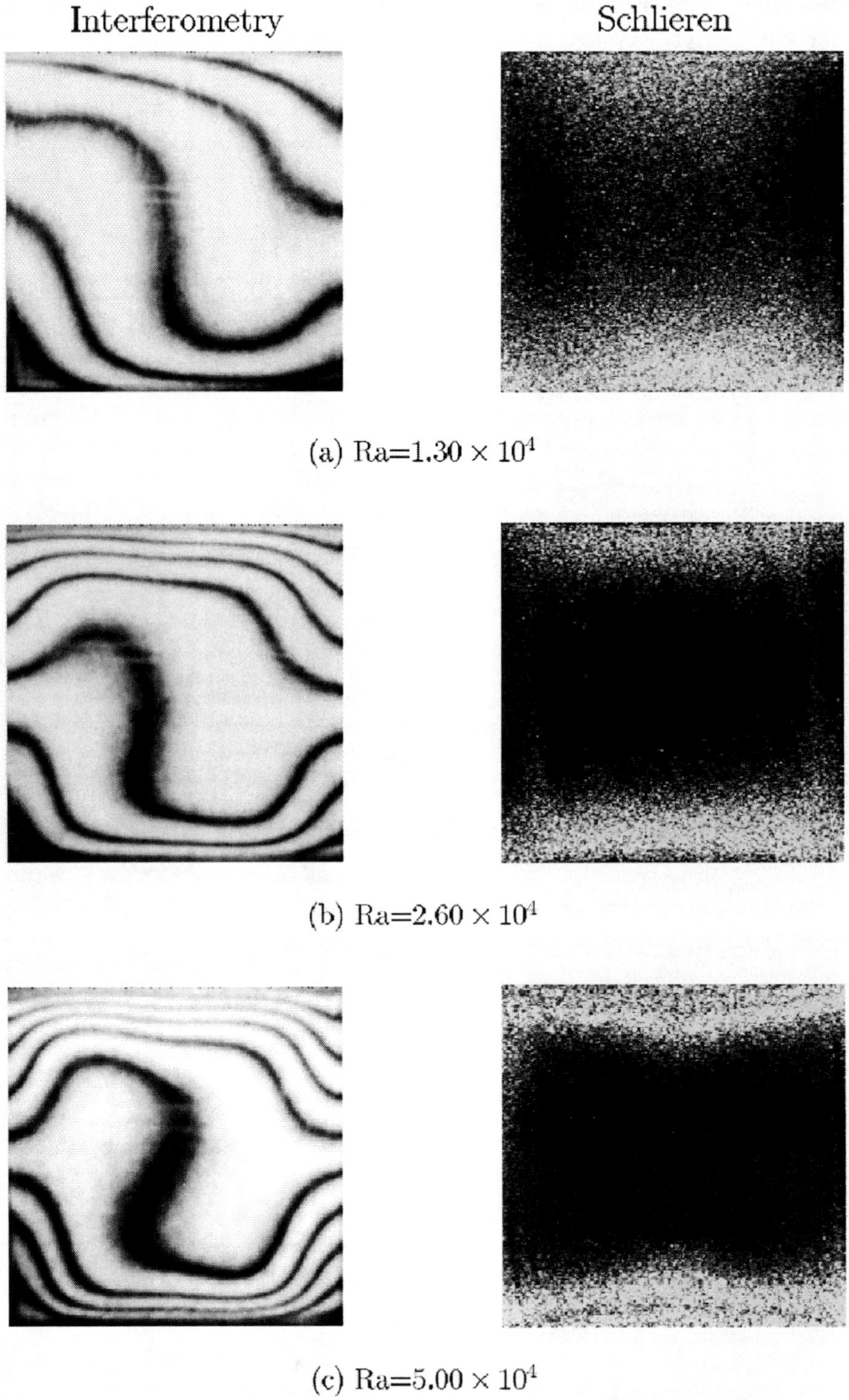

Figure 10. Interferometric and schlieren images for the lower range of Rayleigh numbers in air.

Quantitative analysis of the temperature field and wall heat transfer rates is reported for the lower range of cavity temperature differences (and hence Rayleigh number). Clear shadowgraph images could not be recorded for small temperature differences and have not been shown. At higher temperature differences, the field was seen to become unsteady. The discussion for larger cavity temperature differences is based on qualitative comparison of the three imaging techniques.

Low Rayleigh numbers: Figure 10 shows the steady state interferometric and schlieren images for the lower range of Rayleigh numbers namely $1.3\times10^4<Ra<5\times10^4$. At Ra= 1.3×10^4, the interferogram has only a few fringes. The number of fringes in the experiments was uniformly found to be consistent with the estimate $(T_h\text{-}T_c)/\Delta T_\varepsilon$, the denominator being given by Equation 1. The number of fringes increases with Rayleigh number, along with the fringe density near the horizontal walls. With respect to the schlieren images, it can be seen that the increase in light intensity is distributed over the cavity cross-section at the lowest Rayleigh number. As Rayleigh number increases, the brightness is limited to the wall region, and its size progressively diminishes. The schlieren image clearly brings out a boundary-layer type of flow structure in the cavity. The above experiments have been validated against a numerical model and the comparison has been found to be good [70].

A comparison of the steady dimensionless temperature profiles in the cavity as obtained from interferometry, schlieren, and numerical simulation is presented in Figure 11. Temperature profiles at two column locations (x/h = 1/4 and 3/4) have been considered. The comparison has been presented for the three Rayleigh numbers referred in Figure 10. The shape of the temperature profile, characteristic of buoyancy-driven convection in a Rayleigh-Benard configuration is reflected in all the three approaches. The slopes of the individual curves near the walls give a measure of the wall heat flux. The comparison between the experiments and simulation is seen to be good. Schlieren measurements compare marginally better with numerical simulation, as against interferometry. This is because in interferograms, information about the thermal field is available only at the fringes. Constructing a complete temperature profile requires interpolation between fringes, and is a source of error. In addition, the number of fringes at low Rayleigh numbers is small.

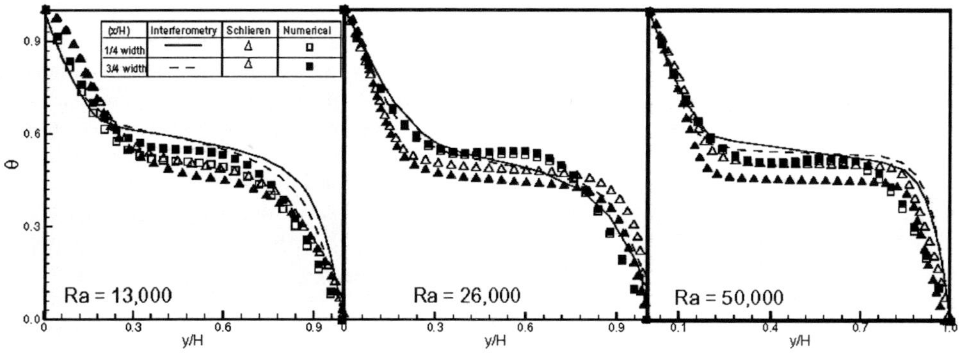

Figure 11. Convection in an air-filled cavity. Non-dimensional temperature profiles as a function of the vertical coordinate.

Under steady conditions, the average wall heat flux is also equal to the energy transferred across any horizontal plane of the cavity. Average heat transfer rates at the hot and the cold walls of the cavity have been calculated in terms of Nusselt number and presented in Table 2. Further, they have been compared against the experimental correlation of [2]. The correlation summarizes the observations of a number of authors and cavity dimensions. Consequently, it has a large uncertainty level of ±20%. The average Nusselt numbers predicted by the two optical techniques match well with each other, with a maximum difference of ±5%, that occurs at the lowest Rayleigh number (Ra= 1.4×10^4) of the present work. The hot and cold wall Nusselt numbers are also reasonably close, indicating a good energy balance in the data reduction procedure. The discrepancies are higher in interferometry at this Rayleigh number owing to the formation of just a few fringes in the cavity.

Table 2. Wall-averaged Nusselt numbers at the lower (hot) and the upper (cold) walls for convection in a rectangular cavity with air as the working fluid. Comparison of experiments with correlation of [2]. Symbol I in brackets indicates values calculated from interferograms; symbol S indicates values from a schlieren image

Rayleigh number	Nu (cold)	Nu (hot)	Nu ([2])
1.4×10^4	2.38 (I) 2.28(S)	1.965(I) 2.07(S)	2.615
2.7×10^4	3.32(I) 3.50(S)	3.13(I) 3.10(S)	3.03
5.1×10^4	3.56(I) 3.37(S)	3.51(I) 3.38(S)	3.45

High Rayleigh numbers: Convection in a cavity subjected to a large temperature difference and hence a high Rayleigh number is discussed in the present section. The shadowgraph images were quite clear in these experiments and have been included for comparison. In each of the experiments, a true steady state was not attained even after the passage of a long period of time. Secondly, interferograms were subjected to large refraction errors. The cavity viewed through the schlieren and shadowgraph layouts looked deformed once again due to refraction. Hence, the present discussion is purely descriptive and quantitative data has not been reported. Rayleigh numbers considered are Ra=8.5×10^4, 1.13×10^5 and 1.4×10^5.

Figure 12 shows representative images of the convective field recorded after the passage of 8-9 hours of real time. At Ra=8.5×10^4, the interferogram shows nearly straight dense fringes parallel to the horizontal walls. These are indicative of near-parallel flow near the hot and the cold walls. The flow turns at the corners, leading further to fringe curvature and separation. The flow field is symmetric about the centerline of the cavity. These trends are well-reproduced by schlieren and shadowgraph images.

The movement of the fluid medium in the cavity cross-section has the shape of a roll. The roll size can be determined directly from the optical images since the fringes as well as the intensity fields should turn along at corners with the local velocity vector. The roll sizes of Figure 12 have been compared in Table 3, and are found to be quite close. The three experimental techniques reveal a reduction in the roll size with increasing Rayleigh number. The nature of unsteadiness in the convective field as recorded by the three optical techniques has been discussed by the authors elsewhere [70].

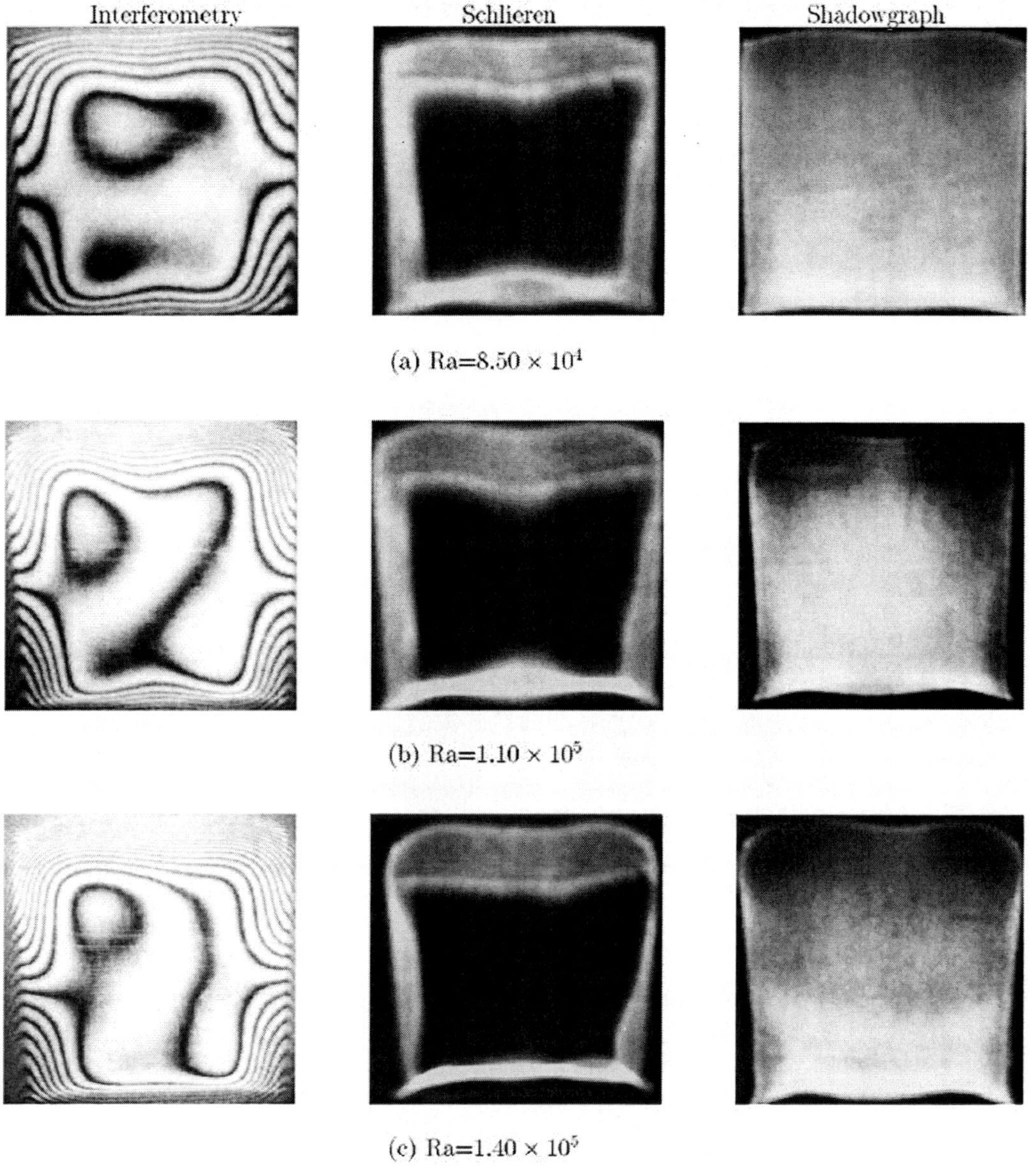

Figure 12. Interferometric, schlieren and shadowgraph experimental images for the higher range of Rayleigh numbers in air.

Table 3. A comparison of roll sizes of convection seen relative to the cavity height in the three optical techniques; effect of increasing Rayleigh number

Rayleigh number	Interferometry	Schlieren	Shadowgraph
8.50×10^4	0.270	0.278	0.310
1.10×10^5	0.258	0.252	0.252
1.40×10^5	0.210	0.242	0.220

Thermal imaging of the test cell filled with water is reported in the following discussion. The vast difference in the properties of air and water leads to certain difficulties in temperature measurement in a water-filled cavity. Specifically, the following issues are to be noted:

1. Even for modest temperature differences, the equivalent Rayleigh numbers are very high. Consequently, the convective field reaches the unsteady, turbulent regime quite early.
2. The temperature drop per fringe shift is small in water. Thus the fringe density in interferograms is high, making data analysis difficult.
3. The sensitivity of the refractive index to temperature is high. Thus the deformation of the cavity size as seen on the screen is large and leads to ambiguity in data analysis.

In view of these complications, it is natural that shadowgraph should be best suited for temperature measurement in water. A comparison of interferometry, schlieren and shadowgraph for a temperature difference of 4 K (Ra=2.5×10^6) is presented in Figure 13. The optical images contain information that is path-integrated in the direction of propagation of light, but are functions of time. Figure 13 shows that the shadowgraph has the most well-defined intensity variation and is the easiest to analyze for the present experiment.

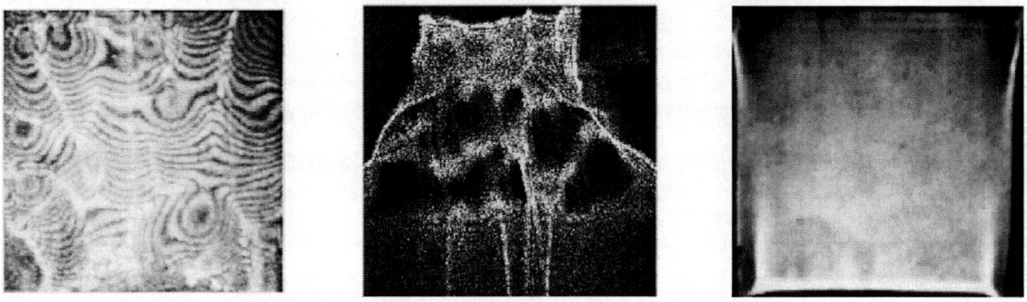

(a) (b) (c)

Figure 13. Comparison of (a) interferometry, (b) schlieren and (c) shadowgraph for buoyant convection in a water-filled cavity at Ra=2.5×10^6.

A comparison of shadowgraph images over a range of high Rayleigh numbers in water is shown in Figure 14. The light intensity variation shows the flow evolving from a boundary-layer type behavior at the lowest Rayleigh number towards large scale structures (namely, plumes) at the highest Rayleigh number. Beyond Ra=5×10^6 the flow is unsteady as well. The images shown in Figure 14 have been time-averaged over a few seconds so that the dominant spatial features are captured.

Convection in water at high Rayleigh numbers in the presence of an air-water interface has been studied by Punjabi *et al.* [68]. A sequence of shadowgraph images recorded in the experiments is shown in Figure 15.

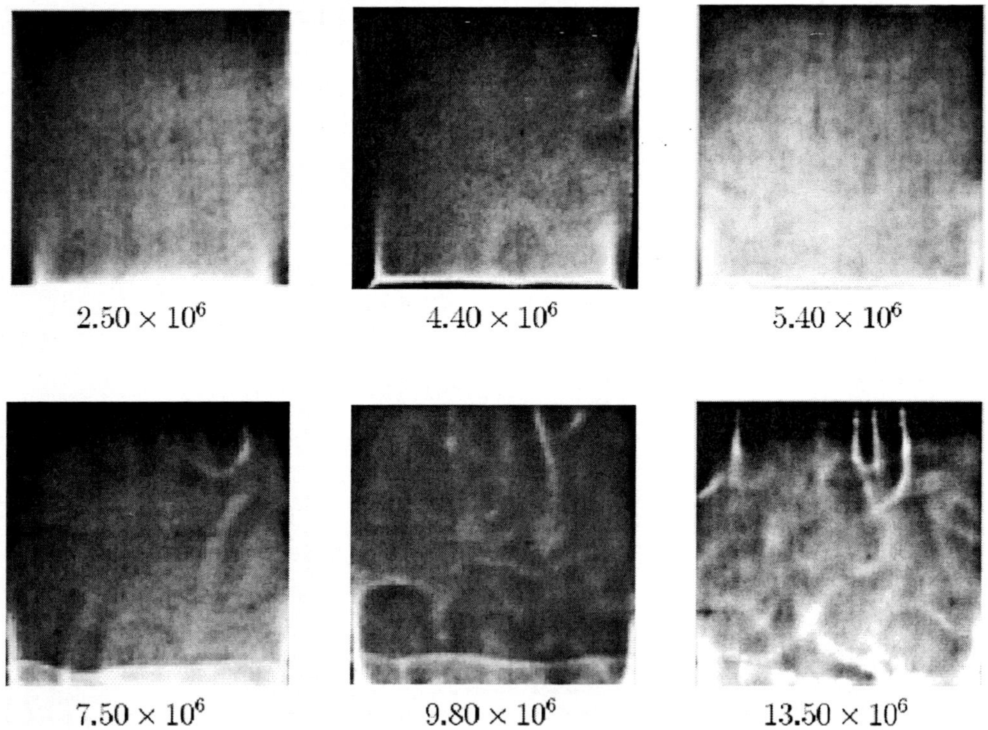

Figure 14. Shadowgraph images of the convective field with increasing Rayleigh number in water.

Based on the above discussion, the following conclusions can be drawn.

1. In low temperature gradient experiments, all the three techniques correlate well with one another. Interferograms are limited by few fringes in air, and too many fringes in water. The shadowgraph image does not show sufficient contrast for analysis. In this respect, the schlieren technique is most amenable to data reduction.
2. In high gradient experiments, both schlieren and shadowgraph yield clear images. The interferograms are however corrupted by refraction errors. Schlieren and shadowgraph track the temporal response of the fluid medium in the form of the light intensity variation.
3. In high Rayleigh number experiments with water, the flow field is turbulent. Shadowgraph images are seen to be meaningful, as against interferograms and schlieren. The shadowgraph images reveal a considerable amount of physical information, including boundary-layers, plumes and time scales. It can be independently used to improve existing models of engineering turbulence.

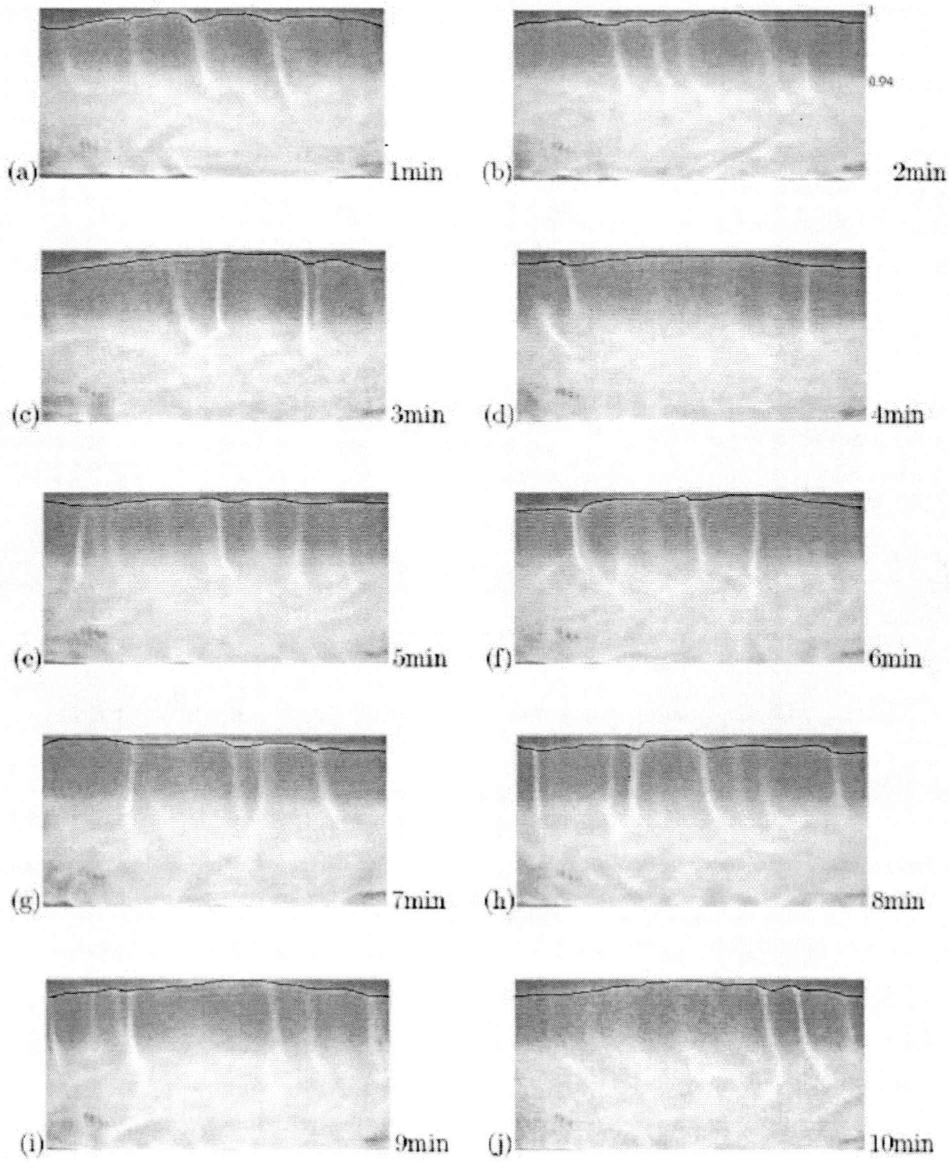

Figure 15. Shadowgraph images showing the deformed interfaces in a cavity half-filled with water, the rest being air; Interfaces are emphasized in black.

9. Convection Phenomena during Crystal Growth

For a purely buoyancy-driven growth process, the driving potential of flow is the maximum concentration difference occurring in the solution. In addition, the strength of convection is also governed by the length scale, typically the size of the growing crystal. With time, the solution is depleted of the salt, though there is an increase in the length scale related to the change in size of the crystal. Jointly, the strength of convection can increase with time, till the solution in the growth chamber is fully depleted of salt. This state is characterized by stable stratification of the solution, the convection currents diminishing in strength to negligible levels. The growth of the crystal practically stops at this stage. Growth can be resumed when the grown crystal is immersed in a fresh supersaturated solution and the ramp rates are re-introduced. The resulting convection patterns would be different from the first stage because of a change in the crystal size. When the crystal is imparted rotation, velocities are created in the angular direction in the horizontal plane, in addition to the buoyancy-driven motion in the vertical plane. However, the two motions are interlinked through the radial component of velocity. The linkage is such that rotational motion leads to Coriolis forces that re-direct fluid motion. However, homogenization of the solution is the dominant factor that reduces concentration gradients, diminishes the driving potential and hence suppresses fluid motion arising from buoyancy. The critical speed at which buoyancy is practically suppressed will depend on the crystal size and overall concentration difference available in the solution. The resulting solutal concentration distribution at the surface of the crystal influences the growth rate and quality.

The process of solute deposition leading to crystal growth occurs on a hierarchy of length and time scales. At the small scale, solute particles arrange themselves as a part of the crystal structure. The pyramidal structure seen at later stages of growth is initiated at this point. The experiments conducted in the present work do not yield information on this aspect of the growth process. At the larger scale (the length scale of the crystal itself), concentration gradients are set up that feed solute to the crystal. These gradients naturally control the rate of crystal growth. The uniformity in distribution of the concentration gradients determines the crystal quality. The present research aims at investigating physical mechanisms at the scale of the crystal in terms of the solutal concentration distribution.

9.1. Comparison of Interferometry, Schlieren and Shadow-Graph in a Crystal Growth Experiment

The applications of interferometry, schlieren and shadowgraph as tools for visualization of buoyant convection and on-line monitoring of KDP crystal growth from its aqueous solution are discussed in the present section. The ramp rate here is around 0.2°C/hour that is quite high in comparison to the rates employed in Section 9.2. The resulting convection patterns, as a result, form earlier in the experiments discussed. Images of the time-evolution of the convective field in the growth chamber have been recorded. The three measurement techniques are compared in terms of image quality and the potential for extracting quantitative information.

9.1.1. Convection Patterns

When a seed of the crystal is inserted in the solution, the initial temperature difference between the two leads to the dissolution of the seed itself. The local density of the solution increases, and consequently the solution descends vertically from the crystal. With the passage of time, thermal equilibrium is established, and density differences within the solution are solely due to concentration differences. Adjacent to the crystal, the solute deposits on the crystal faces, and the solution goes from the supersaturated to the saturated state. Thus the solution near the crystal is lighter than the solution away from it. The denser solution displaces the lighter fluid, and a circulation pattern is set up around the crystal. The structured movement of the fluid, called a buoyant plume is essential for transporting the solute from the bulk of the solution to the crystal and determines the crystal growth rate. The plume is visible in the three visualization techniques: as fringe displacement in interferometry, and the spread of light intensity in schlieren and shadowgraph.

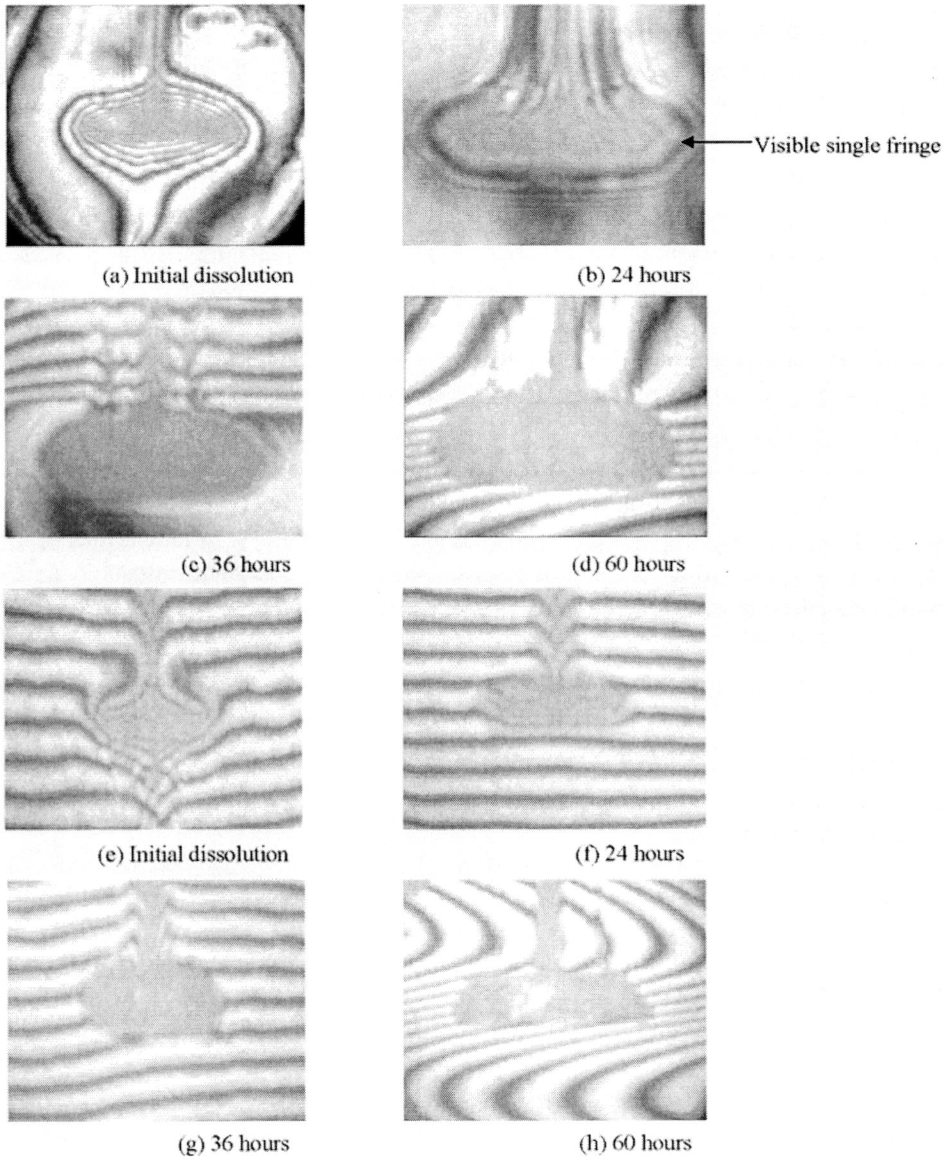

Figure 16. Time sequence of the evolution of interferograms around the growing crystal. (a-d) Infinite fringe setting; (e-h) Wedge fringe setting. The initial crystal size in the infinite fringe setting is greater than in the wedge fringe setting. Large fringe slopes in (e-h) very close to the glass rod are possibly distortions. The opposing fringe curvatures above and below the crystal in (h) show a lighter and a denser solution formed by stratification.

Figure 16 shows the formation and time-evolution of fringes in the infinite and wedge fringe settings of the interferometer around the growing crystal. The first image of the transient shows the appearance of fringes due to crystal dissolution. This process increases the local density, causing the solution to descend vertically downwards. Hence, the fringe displacement is also in the downward direction (Figure 16(a)). Large concentration gradients in the vicinity of the dissolving crystal give rise to time-dependent fluid motion as well. A second factor contributing to unsteadiness is the change in the crystal geometry from rectangular to prismatic. As the crystal attains thermal equilibrium on one hand and its natural shape on the other, the concentration gradients gradually diminish. During a time period of 15-35 hours, the gradients were small enough to produce only a single visible fringe adjacent to the growing crystal (Figure 16(b)). The fringe was found to be stable with respect to time indicating uniform deposition of the solute around the growing crystal. Experiments in the wedge fringe setting of the interferometer reflect identical trends. The horizontal fringes deform vertically downwards (Figure 16(e-f)), and is followed by a phase when they are practically straight. Near the crystal, the solution is practically saturated, while it is supersaturated in the far field. The wedge fringes get displaced in regions of a large change in concentration with respect to the bulk of the supersaturated solution. With further cooling, the solution in the near-field becomes supersaturated with reference to the new temperature, additional salt deposits on the crystal, and the growth process is once again initiated. This leads to concentration gradients adjacent to the crystal, and a continuation of buoyancy-driven flow. In the time frame of 20-30 hours after the insertion of the KDP seed, the infinite and the wedge fringes showed considerable symmetry as well as stability in time, ensuring uniform growth on all the faces of the crystal. This time duration may be called the stable growth regime of the crystal.

With an increase in the crystal size, the influence of even mild concentration gradients is strengthened, increasing the fringe deformation. Over a longer duration of the experimental run time (> 50 hours), the solution is found to be layered (stratified) with respect to density (Figures 16(c-d) in the infinite fringe setting and Figures 16(g-h) in the wedge fringe setting). This is understandable because crystal growth takes place from a fixed volume of the solution in the growth chamber, and with time, the solution is increasingly depleted of salt. The density inversion suppresses convection to a point where the increase in the crystal size is negligible. The downward movement of these layers of constant concentration is driven by molecular diffusion, and contributes to a very slow increase in the crystal size. The appearance of straight horizontal fringes above the crystal in the infinite fringe setting, and opposed curvature of wedge fringes in the far-field can thus be taken as the limiting point where the growth process is to be terminated.

The initial size of the crystal used in the infinite fringe setting had to be larger than that of wedge fringes. This is because the convection patterns with a small crystal did not produce sufficient contrast in the interferograms, and could be recorded by the camera.

The growth sequence using schlieren and shadowgraph techniques is shown in detail in Figures 17 and 18. They confirm the evolution pattern of the convection field recorded by the interferometer. The first image in the schlieren and shadowgraph sequences shows initial dissolution of the seed just after its insertion into the solution. A sharp descending plume originating from the seed can be seen. The intensity contrast is related to an abrupt change in the solute concentration around the seed crystal, which creates a jump in the refractive index, and deflects the light beam into the region of relatively large concentration gradients. The schlieren arrangement is more sensitive to the change in concentration, producing a greater intensity contrast, when compared to the shadowgraph. After the initial dissolution, the growth process of the crystal is initiated. The associated convection currents are larger, and a significant increase in the size of the bright region is seen. The schlieren image is more vivid (Figure 17), when compared to the shadowgraph (Figure 18). The images also reveal the extent of symmetry of the solutal distribution and the underlying flow field in the stable growth regime of the crystal. As in interferograms, schlieren and shadowgraph reveal the following sequence of events: (a) The plumes associated with crystal dissolution are unsteady; (b) the growth process enters a stable growth regime till the crystal size exceeds a critical value; and (c) the flow field approaches a stagnant condition when the solution becomes stratified.

The layering and stable stratification of the solution are seen for times greater than 55 hours (Figures 17(g-h) and Figures 18(g-h)). In the schlieren image, it leads to the region of brightness shifting away from the crystal. Figure 17(h) is an original unprocessed schlieren image. It shows a larger view of convection around the crystal, where stratification leads to a significant refraction and the appearance of a bright patch of light above the crystal. In shadowgraph images, the movement of the stratification front is visible as a bright band that descends vertically downwards. For times greater than 50 hours, the patch of light is localized around the crystal, while alternating bands of bright and dark regions are formed.

The schlieren (Figures 17(c-d)) and shadowgraph images (Figure 18(e-f)) in the stable growth regime show an upward movement of the buoyant plumes around the crystal. These plumes end in the bulk of the solution in the beaker, and descend in such a way as to form a closed loop. The images discussed in the present section pertain to the flow field adjacent to the crystal alone. The buoyant plumes in the stable growth regime are responsible for the deposition of the solute on the crystal surface. These convection plumes are almost steady and uniform in nature resulting in uniform and symmetric growth of the crystal. The gradient of concentration is larger near the crystal surface when compared to the bulk of the solution. This distribution can be visualized as a diffusion boundary layer around the growing faces of the crystal. The thickness of the boundary-layer can be identified as the region over which the light intensity is high. The vertically upward movement of the convective plumes results in the variation of the thickness of concentration boundary layer along the crystal faces. During the growth phase of the crystal, the boundary-layer is the thinnest on the lower side of the crystal, and the highest on the upper side.

The shadowgraph images are not as explicit as the schlieren, but they help in deciding the symmetry pattern of growth, time dependence, and the onset of stratification. It is to be expected that the shadowgraph images would show greater contrast when the gradients are very high, for example crystal growth in forced convection conditions.

9.1.2. Comparison of the Three Techniques

The knowledge of transport phenomena in the growth of crystals from their aqueous solution in the free convection regime is important for understanding the fundamental mechanisms involved and for fixing process parameters. As opposed to forced convection, the free convection technique has its own importance, for example in protein crystal growth where it is the only choice because of their delicate structure and the ease of managing defects. The creation of a homogeneous concentration field in the boundary-layer adjacent to the solution-crystal interface and uniform distribution of solute in the bulk solution are two major requirements. Refractive index-based optical techniques can be used to examine the nature of convection patterns as a function of time and hence the quality of the grown crystal.

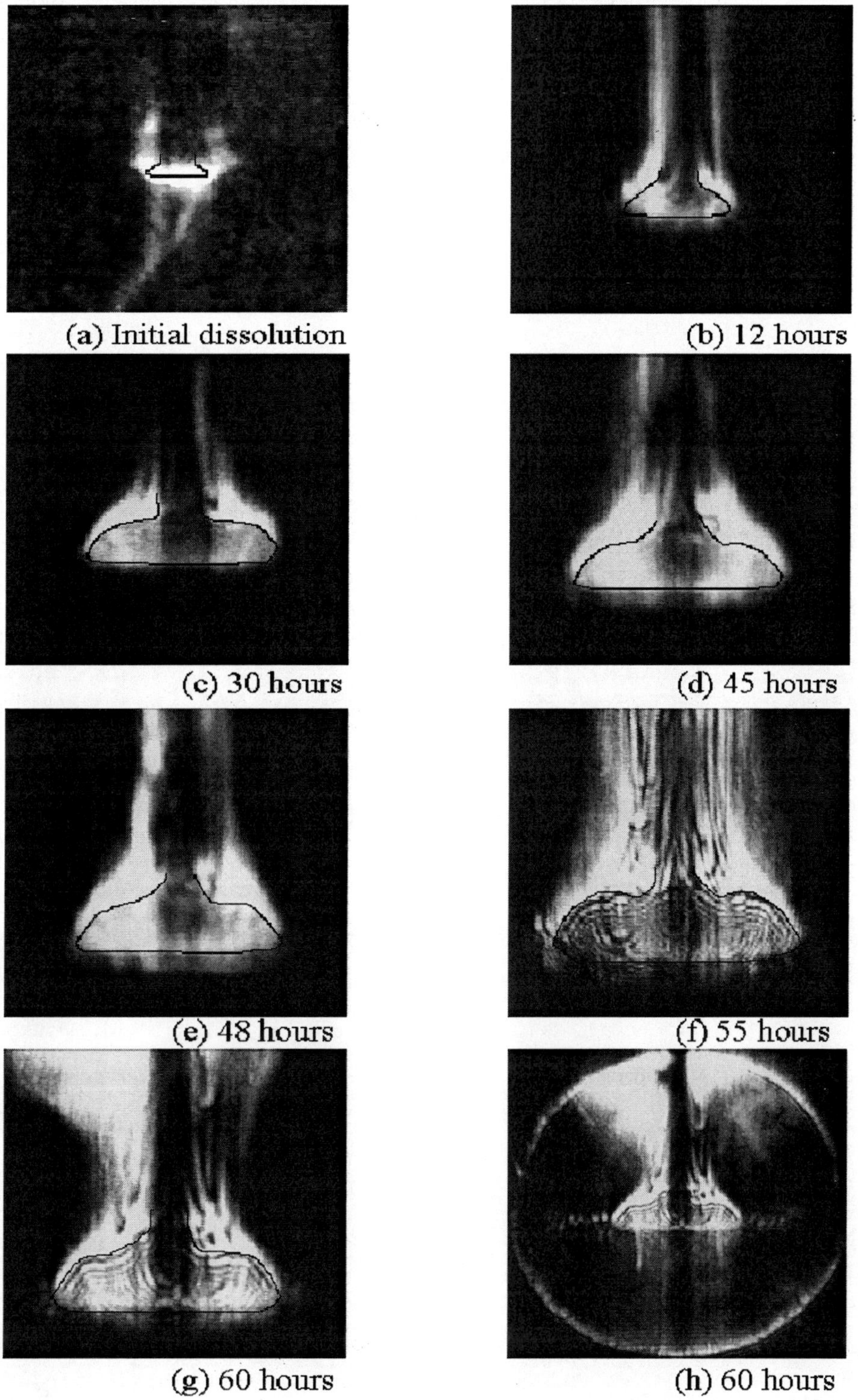

Figure 17. Evolution of schlieren images around the growing crystal from an aqueous solution. Images have been contrast-enhanced to reveal clearly the regions of high brightness. In (h) the original photograph as recorded by the camera is shown. The crystal position has been highlighted.

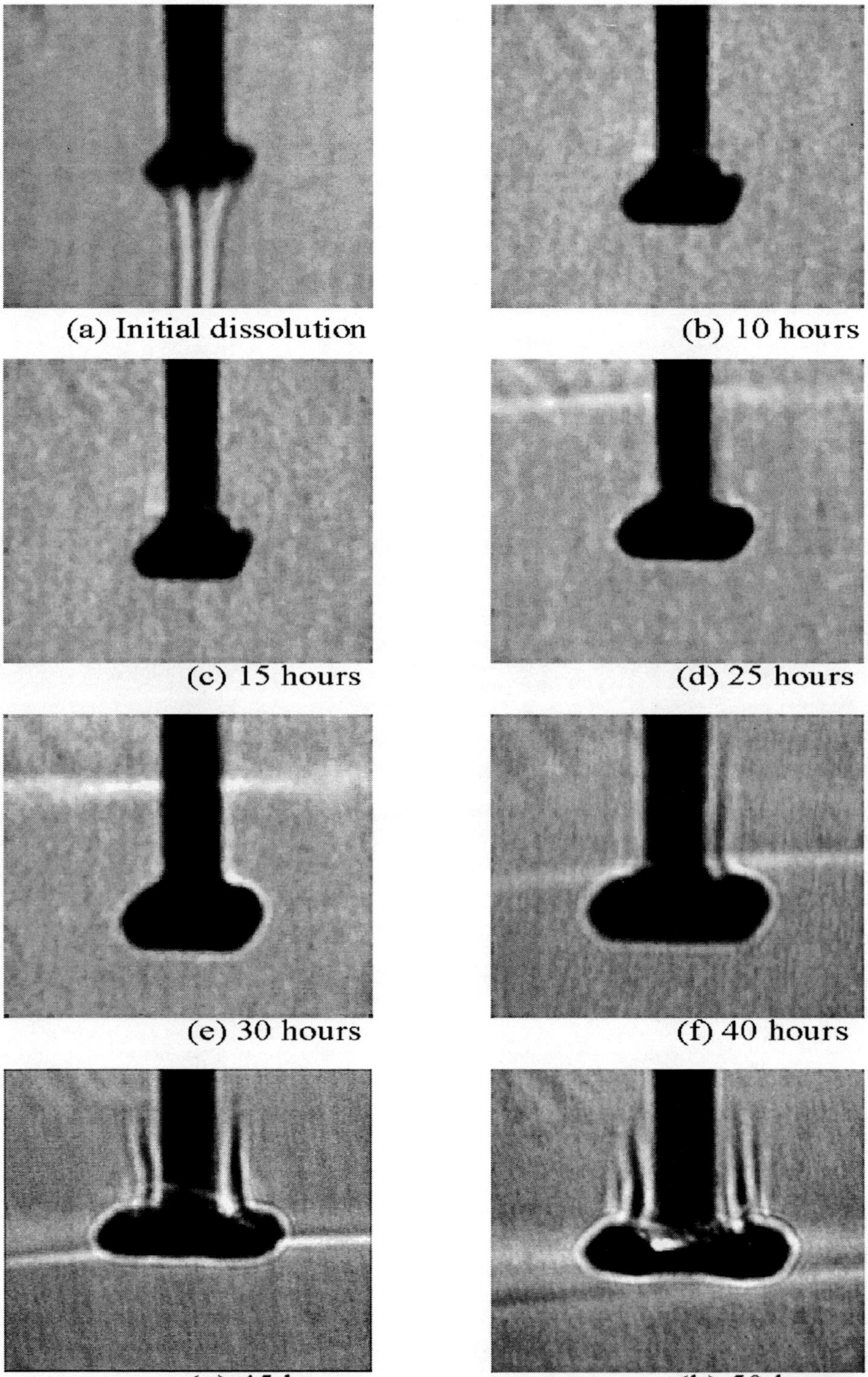

Figure 18. Evolution of shadowgraph images around the growing crystal from an aqueous solution. Images have been contrast-improved for clarity. A bright streak of light indicates the separation of the light solution from the heavy. The streak is seen to move downwards in (d-h), till it stabilizes just around the crystal.

A comparison of the images of the three techniques shows interferograms to be most vivid, since fringes deform and get displaced in relationship to the local velocity field. Thus, they offer the most direct information about concentration distribution as well as the underlying flow field in the solution. Schlieren and shadowgraph images reveal regions of high concentration gradients in the form of heightened brightness, though the former shows greater sensitivity. A review of Equations 1, 3 and 5 (Section 5.3) shows that interferograms are easy to analyze, schlieren requires integration of the intensity field, while shadowgraph requires the solution of a Poisson equation to recover the local concentration.

The three optical techniques under discussion yield images that are integrated values of the concentration field in the direction of propagation of the light beam. Thus, if the spatial extent of the disturbed zone in the solution is small, the information contained in the image is small. In the context of interferometry, the consequence could be the appearance of too few fringes in the infinite fringe setting and small fringe deformation in the wedge fringe setting. In schlieren and shadowgraph, weak disturbances show up as small changes in intensity and hence contrast. The difficulty can be alleviated in schlieren by using large focal length optics so that small deflections are amplified. In shadowgraph, image quality can be improved by moving the screen away from the test cell. Additional difficulties with interferometry are the need for maintaining identical experimental conditions in the crystal growth and the compensation chambers, careful balancing of the test and the reference beams, and limitations arising from fact that quantitative information is localized at the fringes. This discussion shows that configuring the interferometer as the instrument for on-line process control poses the greatest challenge, schlieren and shadowgraph being relatively simpler. Based on the above discussion, schlieren may be considered as an optimum while comparing the ease of analysis with the difficulty of instrumentation.

The quantitative description of the concentration field derived from the schlieren images is discussed in the following section.

9.2. Influence of Ramp Rate and Crystal Rotation on Convection Patterns

The present section discusses the application of the laser schlieren technique to monitoring convection in a crystal growth process from its aqueous solution, when the process parameters are varied. The cooling rate of the solution determines the amount of excess salt available in the solution for deposition on the crystal, and hence the potential difference for driving the convection currents. Ramp rates of 0.05 and 0.1°C/hour have been studied in the present work. These values are smaller than the ramp rate of Section 9.1; consequently the stratification of the solution seen in those experiments was considerably delayed. The rpm of rotational motion fixes the degree of homogenization of the solution and hence, indicates a reduction in the strength of buoyant convection. Crystal rpm of 0 and 15 are studied through experiments. These values have been selected on the basis of their ability to permit growth of crystals of meaningful quality. The effects of ramp rate of the solution, crystal rotation and the size of the growing crystal have been correlated with the growth rate of the crystal. Results of the transient evolution of the convective field in the growth chamber in the form of two-dimensional schlieren images are reported. The images are quantitatively interpreted in terms of concentration contour maps and concentration gradient profiles. In order to bring out the influence of the process parameters, results have been presented in the following sequence: 1. convection currents at a ramp rate of 0.05°C/hour with rotations of 0 and 15 rpm; 2. convection currents at a ramp rate of 0.1°C/hour with rotations of 0 and 15 rpm; and 3. effect of crystal size.

The nature of fluid motion around the crystal gives rise to *boundary-layers* in the flow and solutal concentration fields. These are relatively thin zones adjacent to the crystal faces where large changes in velocity and concentration take place. Large concentration gradients are revealed in a schlieren image as a brightened region against a darker background. The images discussed in the following sections show that the stable growth regime of crystal growth is accompanied by thin high intensity zones that originate at the crystal surface. Thus, it is clear that fluid motion and transport occur in the bulk of the solution, but are governed by the physical conditions imposed by the crystal. These are 1. a prescription of fluid velocity in terms of crystal rpm, and 2. concentration levels fixed by salt depletion from the aqueous solution.

9.2.1. Ramp Rate of 0.05°C/Hour

Figure 19 and Figure 20 show the transient evolution of the convective field for experiments when the growing crystal is respectively held stationary in the solution and rotated at a constant speed of 15 rpm. The ramp rate to cool the solution in the two sets of experiments is 0.05°C/hour. Insertion of the seed into its supersaturated solution can lead to an instantaneous temperature difference between them, followed by an initial dissolution of the crystal. This phase of the experiment is not included in the figures. With the passage of time, thermal equilibrium is established, and density differences within the solution are solely due to concentration differences. Adjacent to the crystal, the deposition of solute from the solution to the crystal surfaces results in a change of concentration and the solution goes from supersaturated to the saturated state. In the absence of rotation, the denser solution displaces lighter solution in the vicinity of the crystal and a circulation pattern is set up around the crystal. The fluid motion is largely in the vertical plane. With rotation, a radial pressure gradient creates an independent circulation loop that forms an alternative basis of solute movement. Here, the fluid particles around the crystal move in the radial direction, but conservation of mass ensures that vertical velocities be set-up once again. In the purely buoyancy-driven mode (0 rpm; also called *natural convection*), the strength and orientation of the convection currents is determined by the available concentration difference in the solution at any instant of time, and hence the cooling rate. On the other hand, an externally imparted rotation to the growing crystal (called *forced convection*) leads to homogenization of the solution, reduction in concentration gradients and hence a reduction in the strength of convection currents. Except for the initial stages of the growth process (where it is diffusion-dominated), these circulation patterns and their interaction form the basis of the transport of solute from the bulk of the solution to the growing crystal surfaces.

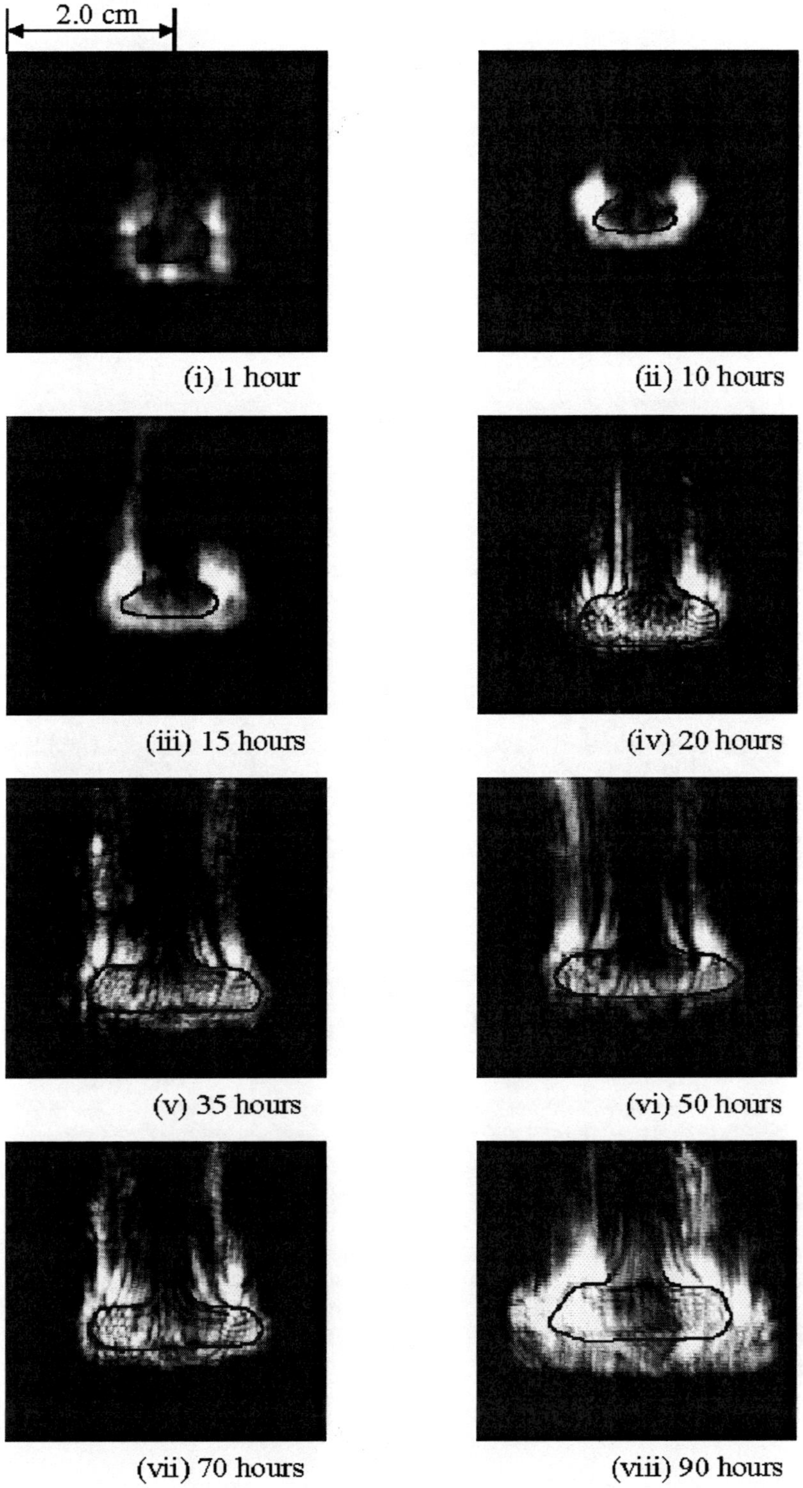

Figure 19. Schlieren images of the transient evolution of the convective field around crystal growing from its aqueous solution. (Ramp rate = 0.05°C/hour, rate of crystal rotation = 0 rpm).

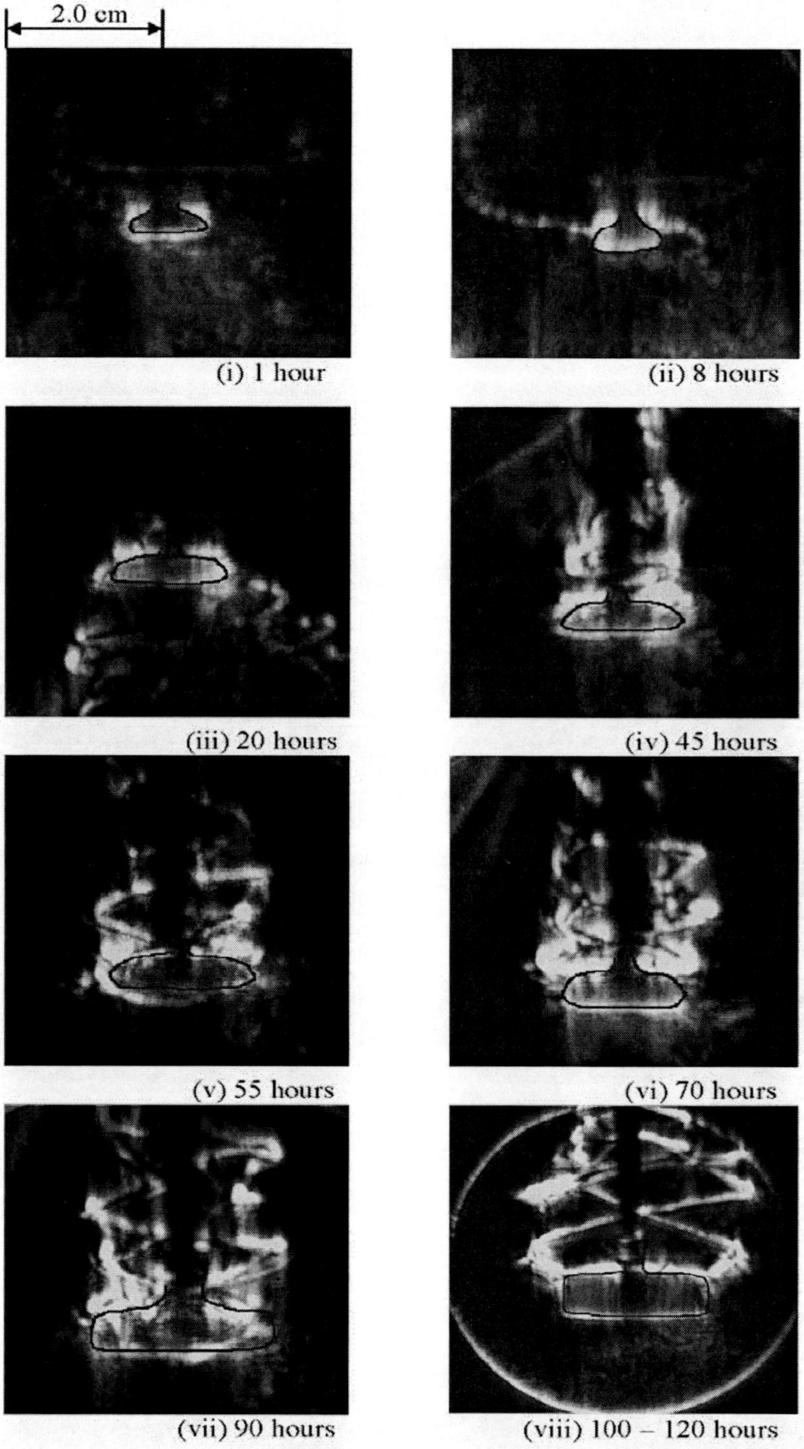

Figure 20. Schlieren images of the transient evolution of the convective field around a rotating crystal growing from its aqueous solution. (Ramp rate=0.05°C/hour, rate of crystal rotation = 15 rpm).

Figure 19 shows the sequence of convection patterns in the purely buoyancy-driven mode. Growth in the initial stages of the experiments is accompanied by steady, weak convection, during which diffusion effects can be expected to be significant. Thus, for $t \leq 20$ hours, a slow growth of the crystal is to be expected. Concentration gradients are primarily localized in the vicinity of the growing crystal. With the passage of time, the size of the crystal increases, and the gradients grow in strength. This result is brought out in the schlieren images as an increase in the light intensity around the crystal. As defined by the bright region, the resulting flow creates a strong plume directly above the growing crystal. Over a longer period of time (20-90 hours), the plume structure remains unchanged. It indicates a stable growth regime for the crystal, where the buoyant plumes are steady and uniform in nature (Figure 19(v-vii)). A gradual evolution of the concentration gradients and the associated buoyant plumes ensure a relatively uniform concentration field in the vicinity of the growing crystal, thus leading to symmetric growth of the crystal at the greatest possible rate. As the crystal increases in size, the convection currents grow in strength. Beyond 90 hours, they are seen to become quite vigorous (Figure 19(viii)). Correspondingly, time-dependent movement of the plumes was seen in the experiments. This stage is characterized by local changes in the concentration gradients in the vicinity of the growing crystal, followed by a breakdown in the symmetry of the growth process. It is a limit on the time duration up to which a single growth experiment can be carried out in free convection regime, and the consequent limit on the size of the grown crystal.

The intensity contrast in the images of Figure 19, as well as their average intensity progressively increases with time, indicating a depletion of the salt in the aqueous solution. For long times, a stable stratification in density (and hence salt concentration) was obtained in the growth chamber [44]. The rate of increase of the crystal size was negligible at this stage.

Figure 20 shows the transient evolution of the convective field around a rotating crystal when the rpm is 15. The first image (Figure 20(i)) shows the existence of a diffusion boundary layer around the surfaces of the seed crystal, as seen by the almost uniform distribution of intensity all over the growing faces. After a growth period of about 10 hours, the gradients near the crystal increase in strength. There is a tendency for the plumes to rise vertically, demonstrating that buoyancy forces are larger than centrifugal, on an average. Unlike the schlieren images in Figure 19 where the buoyant plume moved almost symmetrically along the seed holder over a considerable part of the growth phase, the images shown in Figure 20(ii-iii) reflected temporary unsteadiness in the experiments and asymmetric behavior of the convection currents on either side of the crystal. The unsteadiness can be attributed to two factors: (a) temporary unsteadiness in convection due to the development of prismatic faces from the seed crystal, and (b) the dominance of centrifugal force over the buoyancy force due to crystal rotation, causing the convection currents to be pushed sideways (Figure 20(iii)). At all subsequent time periods, the combined effect of buoyancy and rotational forces govern the overall orientation and movement of the convection currents. As the crystal size increases, relatively stronger convection currents rising upwards due to buoyancy lead to stronger concentration gradients near its surfaces. The effective movement of the fluid particles is along a helical path, seen in Figures 20(iv-vii), but most clearly in Figure 20(vii). The width of the helical structure of the rising plumes scales well with the horizontal dimension of the growing crystal. The increasing strength of buoyant convection is evident from the vertical extent upto which the well-defined helical shape is preserved above the crystal. For example, the respective images for 45 hours (Figure 20(iv)) and 55 hours (Figure 20(v)) show a breakdown of helical structure in the central region between the upper face of the crystal and the free surface of the solution. Helical symmetry is preserved for the greatest vertical extent (i.e., the distance between the crystal and the free surface of the solution) during the time interval of 90-120 hours (Figures 20(vii-viii)). Hence, at a certain level of supersaturation and crystal size, a delicate balance exists between the rotational and buoyancy forces. The balance provides a geometric pattern to the plume, and hence favorable conditions required for the growth of good quality crystals from their aqueous solution is obtained. This particular phase of the growth process can be termed as the stable growth regime in which the crystals of highest transparency and symmetry can be grown. Unlike the growth process in purely buoyancy-driven mode, rotation enforces homogenization of concentration gradients around the crystal over the longer duration of experimental run time. Therefore, the possibility of solute stratification as observed in Figure 19(viii) (and at later times) is delayed.

Figure 20 shows that the convection regime is purely forced for short time, and is governed by crystal rotation (< 8 hours). The stirring effect is to be seen by the streaks of light in the image that spread out deep into the solution. Between 8 and 20 hours, the spread becomes narrower, as buoyancy forces re-direct the plume in the vertical direction. For times greater than 45 hours, the plumes show a swirl component, but are vertically directed. The relative importance of rotation and gravity is governed by the ratio of buoyancy and centrifugal forces. The force ratio can be shown to be proportional to the crystal size [2, 71]; hence buoyancy is the guiding force at later times, when the crystal has become large. However, rotation provides a kinematic condition for fluid motion (in the form of a boundary condition), causing the buoyant plumes to become helical, and hence structured.

Figure 21 shows the concentration contour maps (normalized between 0 and unity) around the growing crystal for 0 and 15 rpm. The value of C=0 represents the saturated state, with C=1 being the supersaturated condition at the temperature of the solution in the growth chamber. The maximum crystal size grown in each experiment has been used to non-dimensionalize the x and y coordinates. In the following sections, the normalized concentration gradient is based on the ratio of the maximum concentration difference and maximum crystal size. The contours in Figure 21(a), corresponding to zero rotation, indicate an almost uniform and symmetric distribution of solute in the growth chamber, both near the crystal and in the bulk of the solution for $t \leq 45$ hours. The contours shown in Figure 21(b) also reflect a similar trend over the comparable time period ($t \leq 55$ hours). Initially (t=1 hour) the contours are localized in the vicinity of the crystal surfaces, with the bulk of the solution being at the supersaturated state (C=1). This phase of the experiment corresponds to the growth process with transport across a diffusion layer, and correlates well with the schlieren images shown in Figure 19 where there is no well-defined upward movement of the buoyant plume. The effect of rotation of the growing crystal can be clearly seen in the contours shown for $1 \leq t \leq 55$ hours, for locations below the crystal (y=0). While those shown in Figure 21(a) (purely buoyancy-driven) depict a larger spread in the upward direction along the seed holder, contours corresponding to 15 rpm are relatively flat. Thus, a uniform distribution of solute is indicated in the solution in the presence of rotation, owing primarily to a mixing action. The vertically rising plume is narrower in the presence of rotation, but broadens with time. At t=45 and 90 hours, Figure 21(a) shows a slight breakdown in symmetry of the concentration contours below the crystal, with respect to the seed holder, though the plume maintains symmetry. Rotation, on the other hand, enforces a better symmetry to the concentration distribution around the crystal. This is clearly visible at 55 hours in Figure 21(b). At 100 hours, the plume is strongly influenced by buoyancy and is as broad as in the case of no rotation. The swirl flow pattern superimposed on buoyant flow by crystal rotation imparts anti-symmetry to the iso-concentration contours. This pattern continues to represent a new form of symmetry for the growth of good quality crystals.

At t=20 hours, an asymmetric set of contours is seen in the region below the growing crystal. This temporary asymmetry in the concentration contours is in agreement with the corresponding schlieren image shown in Figure 20(iii). It is related to the joint influence of buoyancy and rotation. At other times, either rotation or buoyancy predominates, and symmetry in the concentration distribution is restored.

The growth rate of the crystal is related to concentration gradients, rather than concentration alone. This aspect is explored in Figure 22. The variations of the dimensionless concentration gradients averaged over each of the three different faces of the growing crystal (left <001>, right <001> and base <100>) with respect to experimental run time are shown in Figure 22(a) for 0 and 15 rpm. As expected, left-right symmetry of the crystal is realized in growth with and without rotation. The effect of rotation is to lower the overall concentration gradient when compared to that generated by buoyancy alone. The gradients on the lower face are small in comparison to the sides. On the lower face, density stratification is stable and buoyant motion is inhibited. The effect of rotation is then to increase the gradients here by inducing fluid movement. Evolutionary profiles on the top face of the crystal are not shown, since the image is adversely influenced by the presence of the seed holder.

In the purely buoyancy-driven experiment, the gradients on the side faces grow in strength with time, whereas the gradients along the lower face are small. The increase in the gradients on the side faces is consistent with the corresponding high intensity regions in the schlieren image sequence shown in Figure 19. The problem of high concentration gradients during the later stages of experiments ($t \geq 60$ hours) and also a significant difference in the relative distribution of these gradients over sides and lower faces of the growing crystal is seen to be overcome by rotation. The effect of rotation in equalizing the strength of the gradients over the three faces of the crystal is indicated by the proximity of the gradient profiles in Figure 22(a). Figure 22(b) shows the horizontal growth of the crystal with respect to the experimental run time. The growth rate with rotation is slightly lower when compared to that based on buoyancy alone; it is however practically linear. The growth rate with crystal rotation is comparatively lower because of two factors: (a) the lowering of concentration gradients in the vicinity of the growing crystal due to homogenization of the solution induced by crystal rotation, and (b) the rotation of the crystal introduces a radial (outward) velocity component that inhibits the transport of solute to its growing surfaces. Figure 22(c) shows photographs of the grown crystals for the two cases. The size of the finally grown crystal (after 90 hours) is larger in buoyancy-driven convection, but the crystal quality is superior in terms of transparency when growth is accompanied by rotation.

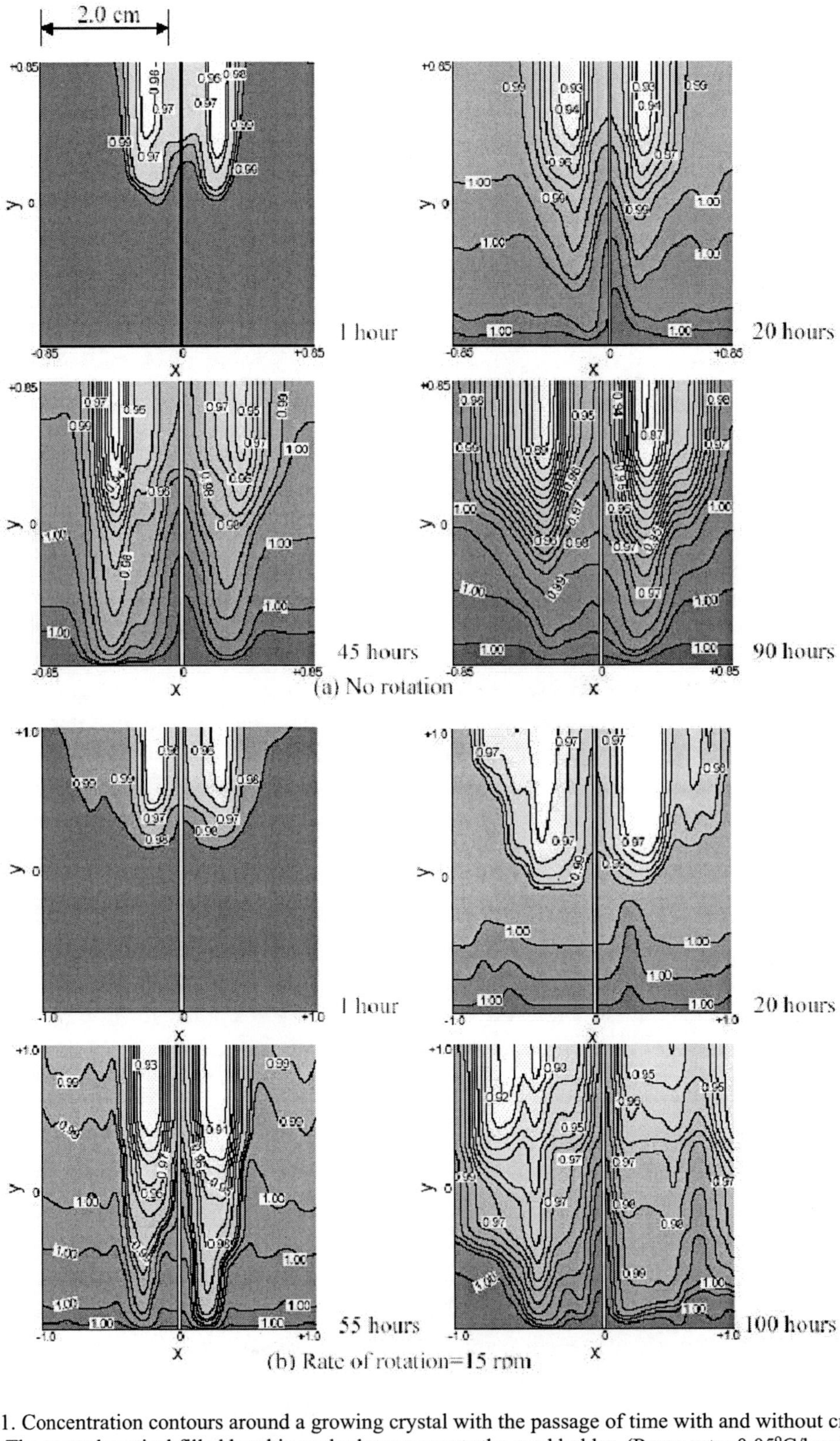

Figure 21. Concentration contours around a growing crystal with the passage of time with and without crystal rotation. The central vertical filled band in each plot represents the seed holder. (Ramp rate=0.05°C/hour).

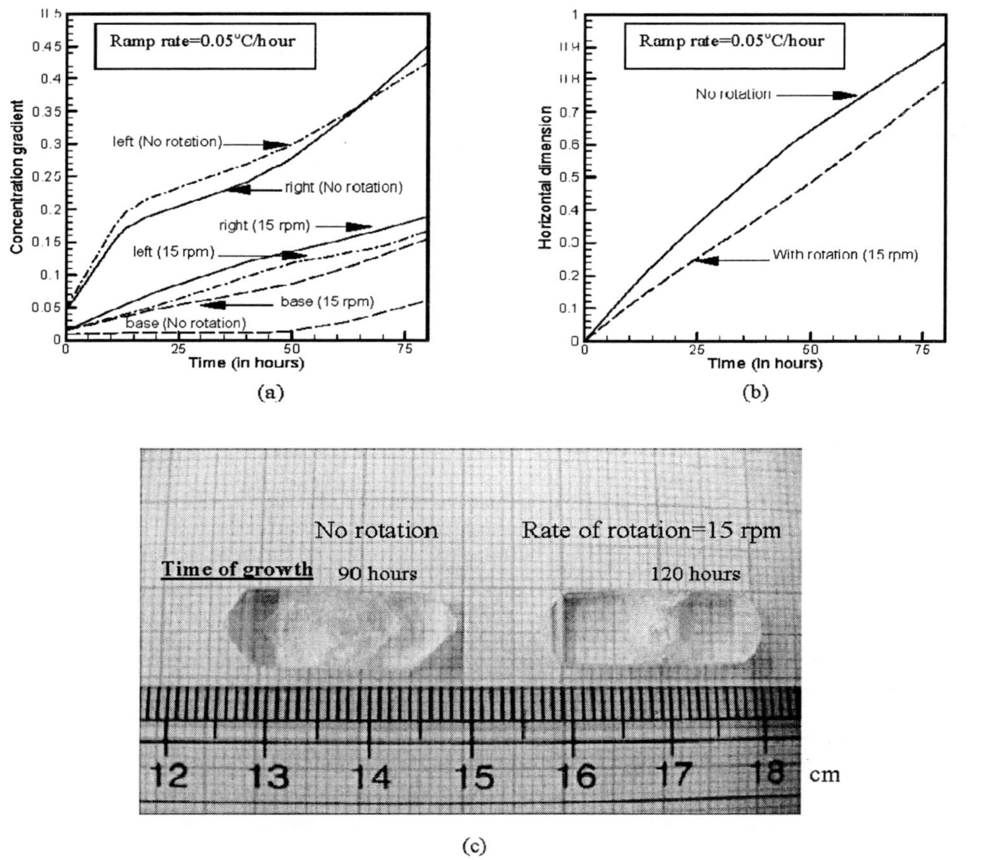

Figure 22. (a) Variation of normalized concentration gradients near the three faces of the growing crystal, (b) horizontal growth of the crystal relative to its maximum size as a function of time, and (c) photographs of grown crystals with and without crystal rotation. (Ramp rate=0.05°C/hour).

9.2.2. Ramp Rate of 0.1°c/Hour

When the ramp rate is increased, the amount of salt that is held in the supersaturated solution increases. The consequence is a greater driving potential for buoyant convection. There is an outer limit to the degree of super-saturation of the solution, though it was not reached in the present study. The growth process of a crystal when the solution is cooled at a higher ramp rate of 0.1°C/hour is discussed in the present section.

Figure 23 shows the transient evolution of the convective field in a purely buoyancy-dominated growth regime (0 rpm, 23(a)) and under the combined effects of buoyancy and rotation of the crystal (15 rpm, 23(b)). In comparison to the sequence shown in Figure 19 (ramp rate=0.05°C/hour, 0 rpm) the region in the vicinity of the growing crystal reveals a bright intensity distribution, indicating larger concentration gradients in the buoyancy-driven regime. The spread of the bright patch at t=1 hour shows that fluid motion has set in quite early. Thus, even the initial growth is not diffusion-controlled. The buoyancy-driven fluid motion is more vigorous at the higher ramp rate, leading to a greater vertical spread of the depleted solution, and a narrower plume. A clear movement of the buoyant plumes on either sides of the seed holder can be seen at t=35 hours in Figure 23(a), whereas at a comparable time instant, weak convection currents were observed at a lower ramp rate of Figure 19.

Schlieren images in Figure 23(a) show well-defined, strong buoyant plumes that transport solute from the bulk of the solution to the crystal surfaces at a faster rate. Consequently, the crystal size at 65 hours in Figure 23(a) is larger than that at 90 hours in Figure 19. Rotation of the growing crystal tends to diminish the high concentration gradients in the growth chamber. It is clearly seen from the schlieren images shown in Figure 23(b) for $t \leq 25$ hours. In contrast to pure buoyancy, the convection currents do not rise vertically upwards; instead they are dispersed in the solution as they are determined by the combined effects of buoyancy and rotational forces (Figure 23(b), t=25 hours). The sizes of the crystal and the convection pattern in the mixed regime (when rotation and buoyancy are of comparable magnitude) are similar in Figures 20 and 23(b). Furthermore, lower concentration gradients due to the homogenization of the solution are reflected by the weak intensity contrast in Figure 23(b) in comparison to that in Figure 23(a).

As the crystal size increases, vigorous convection currents in the solution due to buoyancy result in steep concentration gradients around the crystal (Figure 23(a), t=65 hours). In the presence of rotation (Figure 23(b), t=50 and 80 hours), the plumes orient themselves in the vertical direction, showing that buoyancy predominates over rotation. However, the flow field is clearly disorganized, leading to the appearance of light streaks that are smeared in the horizontal direction by the circulating flow. There is a considerable loss of symmetry in the concentration field at all time instants. At t=80 hours, the higher ramp rate removes a considerable amount of solute from the solution, leading to density stratification in the growth chamber (Figure 23(b)). Rotation has no impact on the concentration field at this stage.

In the presence of rotation, the influences of a lower and a higher ramp rate can be discussed as follows. In the former experiment, rotation imparts symmetry of the solute distribution, diminishes the gradients, the crystal growth rate and results in crystals of better transparency. In the latter, rotation has a less significant role. The solute distribution is homogenized, but there is no improvement in the symmetry of the concentration contours. The crystal growth rates remain high and the solution is density stratified in a shorter period of time. The growth of the crystal is halted at this point. The above discussion shows that for a given ramp rate, a judicious choice of crystal rotation can lead to an enhanced crystal quality, though a slower growth process.

Figure 24(a) shows the evolution of concentration gradients on the faces of the growing crystal for a ramp rate of 0.1°C/hour. In the purely buoyancy-driven growth regime, the gradients on the side faces of the crystal progressively keep increasing. Thus, very high concentration gradients are obtained towards the late stages of growth. On comparing with the corresponding variation of concentration gradients with time in Figure 23(a), the effect of higher ramp rate of the solution can be clearly understood. The maximum non-dimensional value of concentration gradient for the ramp rate of 0.1°C/hour is around 0.55 observed at t=60 hours. For the lower ramp rate (0.05°C/hour), the corresponding value is 0.45 after 80 hours. To some extent, rotation lowers the monotonic increase in the concentration gradients in Figure 24(a), and improves the left-right symmetry. The difference is on the lower side, where rotation further diminishes concentration gradients, in comparison to growth without rotation. This result is the reverse of what is seen in Figure 22(a). The large gradients on the side faces, and low gradients on the horizontal faces leads to high growth rates of the horizontal dimension of the crystal (Figure 24(b)), that is relatively insensitive to rotation. The quality of the grown crystals (Figure 24(c)) is inferior in transparency to that obtained with a lower ramp rate (0.05°C/hour) that is shown in Figure 23(c).

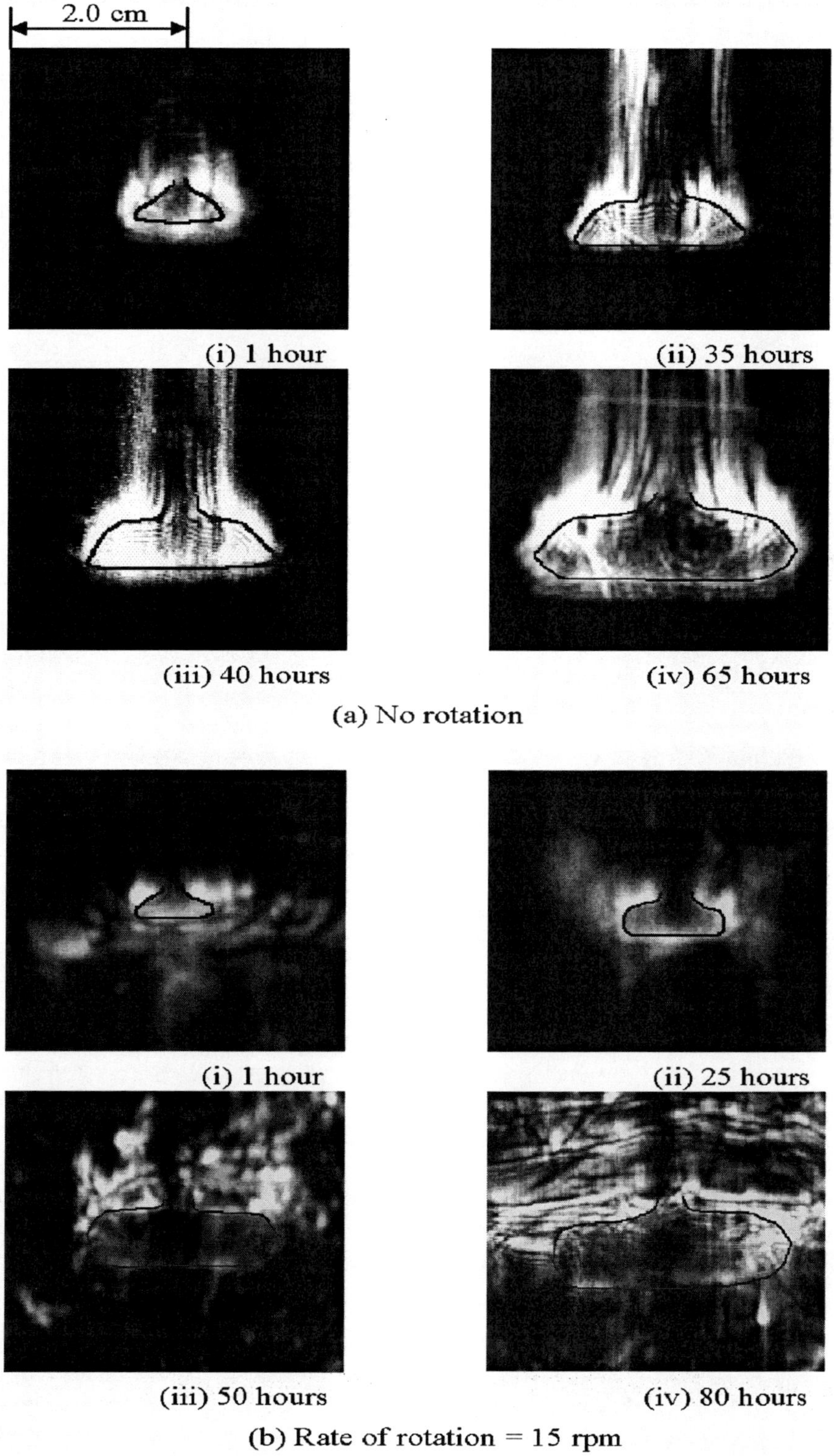

Figure 23. Schlieren images of the transient evolution of the convective field around a KDP crystal growing from its aqueous solution with and without crystal rotation. (Ramp rate = 0.1°C/hour).

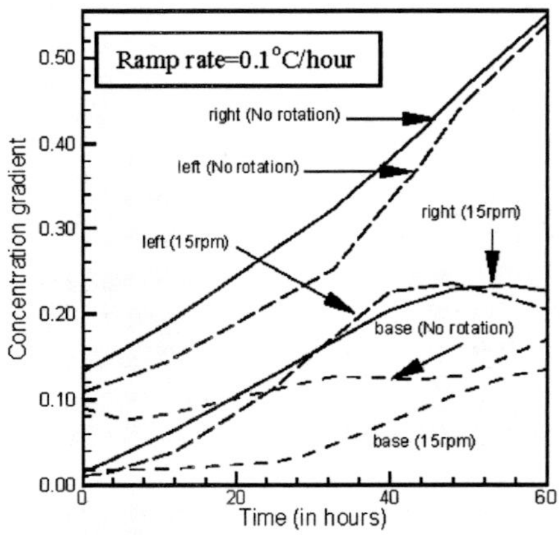

(a)

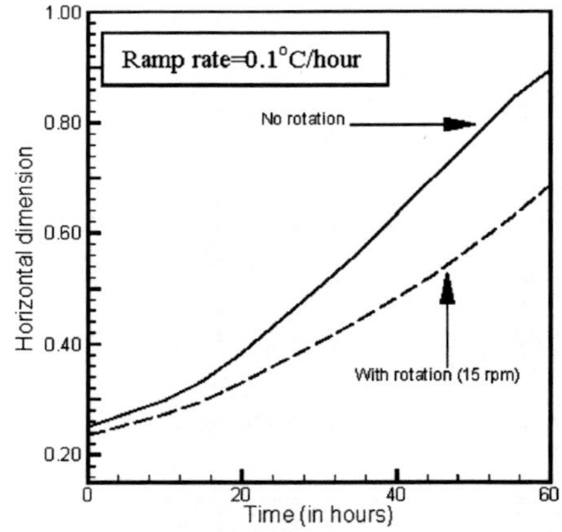

(b)

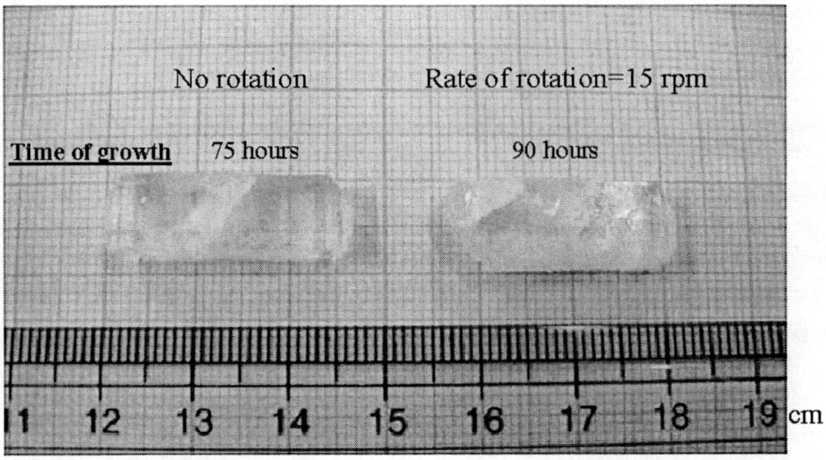

(c)

Figure 24. (a) Variation of normalized concentration gradients near the three faces of the growing crystal, (b) horizontal growth of the crystal relative to its maximum size as a function of time, and (c) photographs of grown crystals with and without crystal rotation. (Ramp rate=0.1°C/hour).

9.2.3. Effect of Crystal Size

In the above sections, it is seen that the crystal growth rate is positively correlated with the local concentration gradients. The gradients in turn depend on the size of the crystal, ramp rate and the rate of rotation. It is of interest to know if the gradients are explicit functions of time. An absence of time-dependence would show the process to be quasi-steady, and the results obtained in the present work would be applicable with greater generality.

To test the experimental data for the presence of growth history and inertia effects, the following experiments were performed: Previously grown crystals of small, medium and large sizes were suddenly inserted into the supersaturated solution. The solution was subjected to cooling at the prescribed ramp rate (of 0.1°C/hour). The short time convection currents around a passive crystal have been compared with those that evolve gradually around a growing one when the growth process takes place from a seed, for a longer duration of time

(~ 60-70 hours). A favorable comparison in terms of convection patterns is an indicator that the concentration gradients respond directly to the process parameters, and are not explicit functions of time.

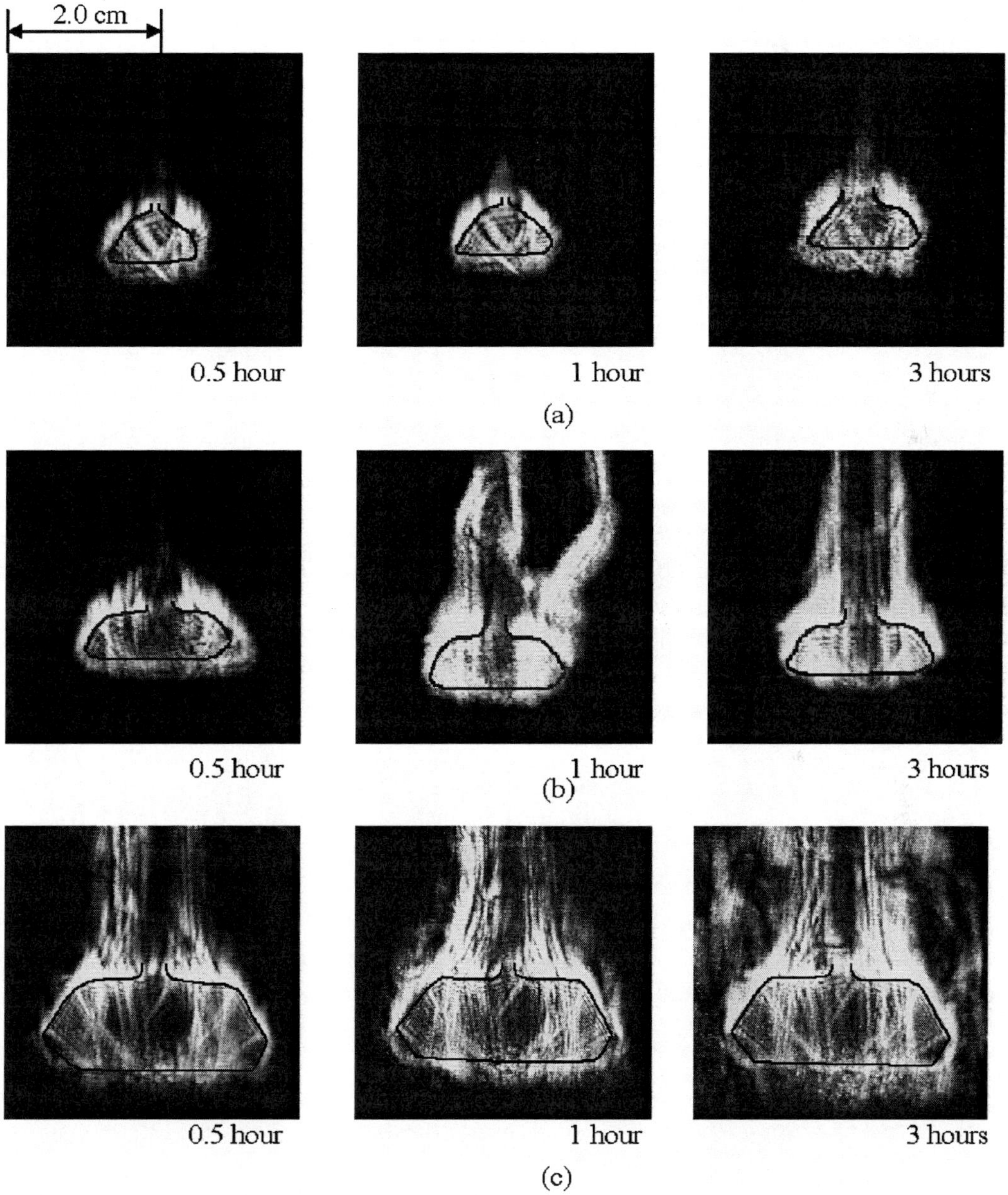

Figure 25. Schlieren images showing the instantaneous evolution of buoyancy-driven convection currents in the growth chamber when a crystal of (a) small, (b) medium and (c) large size is inserted into its supersaturated solution.

Figure 25 shows instantaneous schlieren images of buoyancy-driven convection around the crystal of varying sizes (small, medium and large) over duration of 3 hours. Compared to the time durations generally required, this is a small time scale. Figure 25 shows that short time transients are indeed present in the solution, but they decay within a time frame of three hours for the three sizes considered. Over this time frame, the change in the crystal size was found to be negligible. Figure 25 shows that the strength of convection is weaker during the transient phase, and is the most vigorous when steady state appropriate for the size of the crystal is reached. The convection plumes during the transient phase can be unsteady and unsymmetrical. The loss of symmetry is related to the imperfections in the initial crystal geometry. The unsymmetrical plume is responsible for inducing flow unsteadiness above the crystal. With the passage of time, minor imperfections of the crystal are accommodated and the plume becomes symmetric as well as steady. The schlieren images in Figure 25 recorded at the end of 3 hours of experimentation correlate well with those shown in Figure 23. The similarity of buoyant plumes in the two configurations reveals that the long-time crystal growth process is indeed quasi-steady. Figure 25 also shows that departure from quasi-steady growth conditions can result in loss of symmetry and unsteady plume movement, both of which are undesirable from the view-point of crystal quality.

9.2.4. Competing Mechansims In Convection

The relative importance of buoyant convection over rotation is discussed in the present section in fundamental terms. Essentially, the increase in crystal size takes place owing to the deposition of excess salt in the solution on the crystal faces. The speed at which the deposition occurs is proportional to the gradients in solutal concentration prevailing in this region. The gradients in turn depend on the pattern of fluid motion in the crystal growth chamber. The influence of fluid motion on concentration gradients can be explained in terms of solutal boundary-layers (*diffusion boundary-layers*) adjacent to the crystal faces. The boundary-layer thickness decreases as the fluid velocity increases. Clearly vigorous convective motion will increase concentration gradients, ultimately leading to a rapid increase in crystal size.

The conditions under which flow is intensified can be understood in terms of the forces acting on the fluid particles. For the solution contained in the beaker, the external force arises from buoyancy (in a gravitational field), while inertia forces related to linear and rotational acceleration are also present. The motion, in all situations is retarded by fluid viscosity. The forces are conveniently explained in terms of the dimensionless parameters Ra, the Rayleigh number Ra and Re, the Reynolds number. As discussed in Section 4.1, Rayleigh number denotes the ratio of energy available in a buoyancy field and energy required to overcome viscous forces. Reynolds number is a ratio of rotational and viscous forces. These parameters are respectively defined as [2, 71]

$$\mathrm{Ra}=g\beta \Delta C L^3/\nu\alpha;\ \mathrm{Re}=\omega L^2/\nu$$

Here, g is acceleration due to gravity and β is the volumetric expansion coefficient of the solution with respect to salt concentration (=0.1 per mol). Further ΔC is the concentration difference between the saturated (density 1.1717 gm/cm^3) and super-saturated solution, L the crystal thickness in the vertical direction, ν the kinematic viscosity ($=1.073\times10^{-6}$ m^2/s) and α the mass diffusivity (1.0×10^{-9} m^2/s). The symbol ω is the crystal rotational speed. In the present set of experiments, the linear dimension L increases with time, till the excess solute in the vicinity of the crystal is fully utilized.

The relative importance of the Rayleigh and Reynolds numbers is defined by the ratio $A=Ra/Re^2$, often called the Archimedes number [2]. When the parameter is far away from unity, Ra (for A>>1) or Re (for A<<1) alone will control the flow. For the ratio A close to unity, one can expect an interaction between the buoyant and rotational force fields. It is reasonable to expect that the critical value of the Archimedes number that defines transition from buoyancy to rotation-controlled flow regimes is geometry dependent. Two and three-dimensional geometries are expected to yield different critical values. In the absence of definite information, A=1 is used in the following discussion as a critical value.

For the experiments conducted in the present work, the Rayleigh number based on the smallest crystal diameter is 10^4, while based on the largest crystal diameter it is in excess of 10^6. The corresponding Reynolds numbers range from 0 (no rotation) to 8 (smallest crystal) to 110 (largest crystal). These changes occur primarily because of an increase in the crystal size, but also due to an increase in the concentration difference available in a higher ramp rate experiment. For a small rotating crystal, one can calculate the ratio A=156, while for a large rotating crystal, A=82. Clearly, A>>1 in all the experiments discussed above, and fluid motion is controlled by buoyancy, thus ensuring large-scale circulation in the crystal growth chamber. Rotation plays an important role locally in equalizing solutal concentrations and maintaining stable growth conditions.

The directional nature of gravity imparts certain peculiar properties to the buoyancy field. Specifically, the strength of convection patterns over the horizontal faces of the crystal depends on whether the face is aligned or opposed to the gravity vector. The downward facing surface is stabilized by gravity, experiences weak convection and has the smallest concentration gradients (as seen in Figure 22(a)). At a later point of time, convection in the bulk of the solution drives fluid flow over the lower surface as well, thus increasing the concentration gradients. A second consequence of weak convection is that the influence of rotation is felt strongly even at small time leading to a rapid increase in concentration gradients. This result is once again to be observed in Figure 22(a).

9.3. Tomographic Reconstrcution of Schlieren Data

The question of a correlation between the solutal distribution in the solution and the topography of the crystal can be partially addressed by reconstructing the solute concentration contours on select planes above the crystal. Reconstruction from schlieren images using principles of tomography is discussed in the present section. Unlike the hanging crystal configuration in Sections 9.1 and 9.2, the crystal in the set of experiments was mounted over a platform. With this arrangement, the interference of the rising convection plume with the crystal holder is eliminated. Recording projections requires the apparatus to be turned; in this respect, what is collected is time-lapsed data. For meaningful analysis, it is thus necessary that the convection pattern be steady in time. Only those experiments where this requirement was fulfilled have been included in the discussion.

One parameter that defines the quality of the growing crystal is the symmetry of its faces. Since growth from an aqueous solution is governed by the strength and movement of buoyancy-driven convection currents, the convective field plays an important role in transporting solute from the bulk of the solution to the crystal faces. Hence, to ensure symmetric growth of the crystal, it is necessary to ensure that the convective field in the growth chamber retains a symmetry pattern. In the present work, a viewpoint that an axisymmetric convective field is favorable for crystal growth has been adopted. Thus, conditions under which the solutal distribution is axisymmetric have been delineated from the tomographic reconstructions.

Results are presented for experiments on crystal growth in the diffusion as well as steady convection regimes. In Section 9.4, results have been presented in the form of a time sequence of schlieren images recorded at four view angles, corresponding to four pairs of optical windows of the growth chamber. The crystal is not disturbed during the growth process. In this respect, the measurement process is truly non-invasive, though with a drawback of allowing only a limited number of projections. The emphasis is towards understanding the diffusion-dominated and the stable growth regimes of the crystal. In both regimes, the concentration field is practically steady, particularly during the recording of the individual projections. The reconstruction of the concentration field over selected planes above the crystal and their relationship to the crystal geometry are discussed. The influence of crystal rotation on the convective field is not covered in the present discussion.

9.4. Convection around a Growing KDP Crystal

Imaging of three dimensional convection patterns around the growing KDP crystal is discussed here. Four view angles are considered for tomographic reconstruction of the concentration field. In the experiments conducted, the crystal is not disturbed during the recording of the projection data. In addition, longer durations of crystal growth have been considered. Since finite time is required to turn the beaker and record projections, experiments have been conducted when the concentration field around the crystal is nominally steady. Occasionally, the plume above the crystal is marginally unsteady, and a time-averaged sequence of schlieren images has been used for analysis. The increment in the view angles is 45°, covering the range of 0 to 180°.

The crystal growth process has been initiated by inserting a spontaneously crystallized KDP seed into its supersaturated solution at an average temperature of 35°C. This step is followed by slow cooling of the aqueous solution. The cooling rate employed is 0.05^{o}C/hour. The seed thermally equilibrates with the solution in about 20 minutes. With the passage of time, the density differences within the solution are solely due to concentration differences. Adjacent to the crystal, the solute deposits on the crystal faces, and the solution goes from the supersaturated to the saturated state. When the crystal size is small, salt deposition from the solution on to the crystal occurs by gradient diffusion. Diffusive transport has been sustained for a longer duration of time (and larger crystal sizes as well) by maintaining a small degree of supersaturation in the solution.

9.4.1. Growth Patterns in the Diffusion Regime

Figure 26 shows schlieren images of a growing crystal in the diffusion-dominated regime at three different time instants. All four view angles are included in the figure. Adjacent to the crystal the deposition of solute from the solution to the crystal surfaces results in a change of concentration. This introduces a variation in density, refractive index and hence results in intensity contrast. For small time (less than one hour), intensity changes are restricted to the vicinity of the crystal. Here, transport of the solute to the crystal is purely by diffusion. At later times, a weak convection current is to be seen in the spread of light intensity above the crystal, though diffusion is still the dominant mechanism for solutal transport. For times greater than 70 hours, steady convection was seen to be clearly established in the form of a buoyant plume. In view of the small degree of supersaturation, the origin of convection here is related to an increase in the crystal size, rather than density differences.

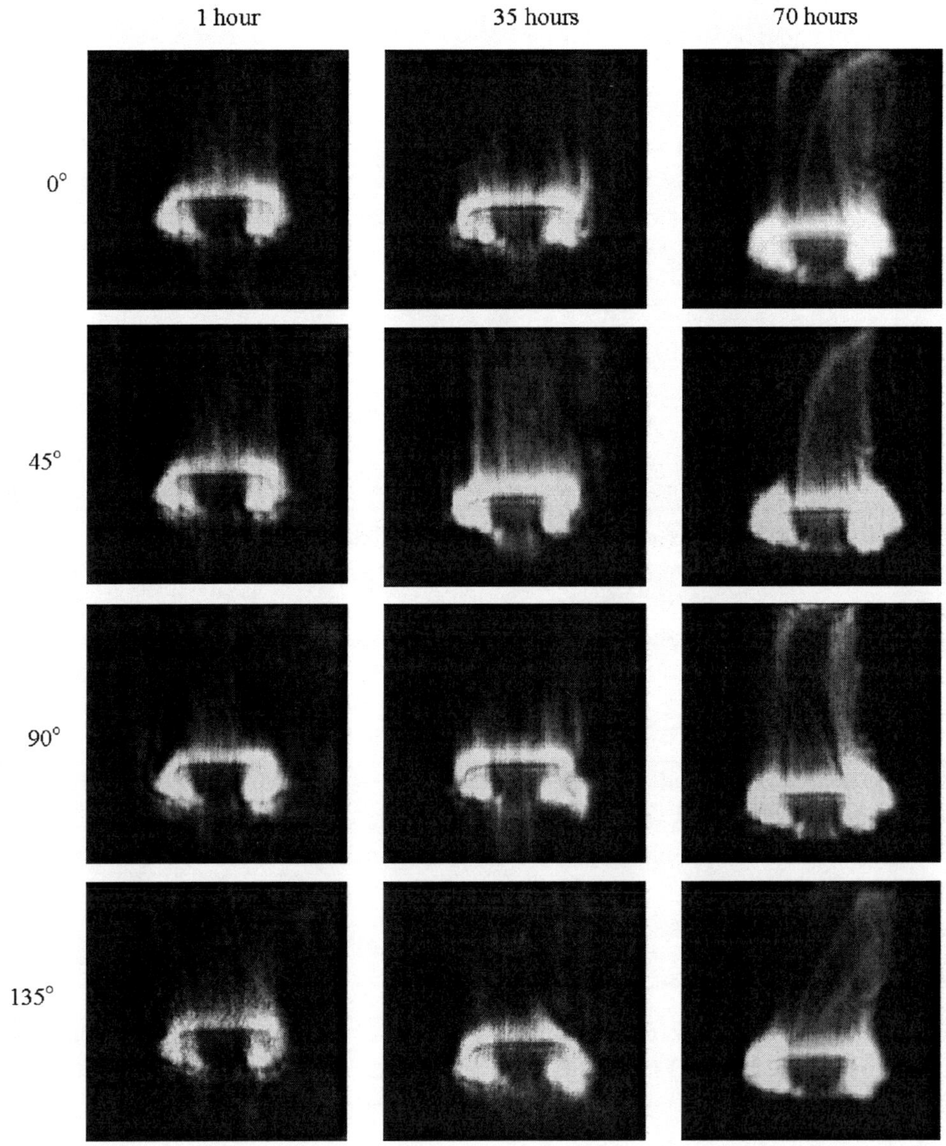

Figure 26. Schlieren images of the convective field around the crystal growing from its aqueous solution in a platform technique at three different time instants recorded from the four view angles are presented.

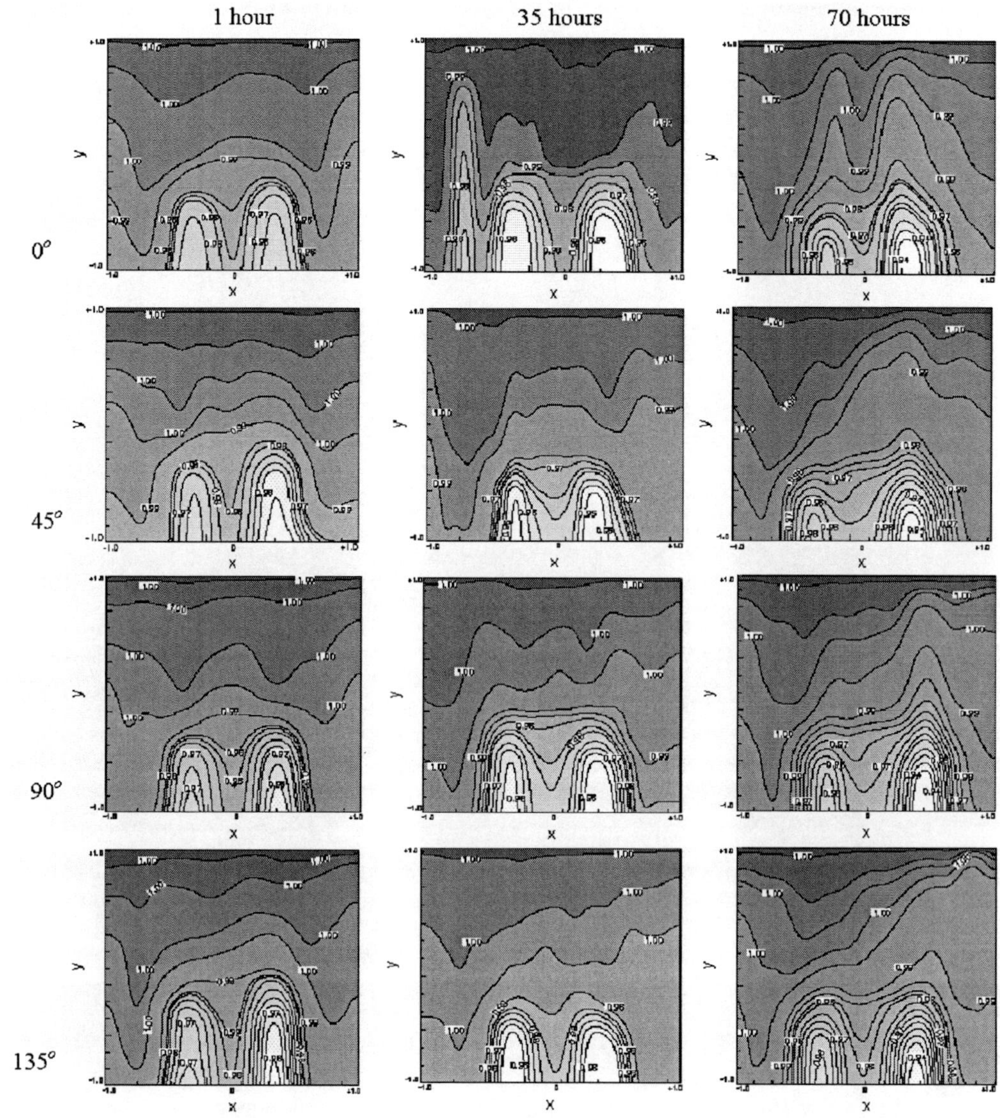

Figure 27. Concentration contours around a growing crystal at three different time instants in a diffusion dominated growth process for the four view angles.

Figure 27 shows normalized concentration maps above the growing crystal as derived from the schlieren images of Figure 26. The value of C=0 represents a saturated solution, while the limit C =1 refers to the supersaturated solution at its instantaneous temperature in the growth chamber. The maximum crystal size at the end of each experiment has been used to non-dimensionalize the x and y coordinates. For t=1 hour, the concentration contours of Figure 27 in the bulk of the solution reveal an almost symmetric solute distribution. Symmetry can also be gauged by the identical nature of contour distributions for each view angle. The contours are not densely spaced, indicating low concentration gradients. The contours are localized in the vicinity of the crystal, with the bulk of the solution being at the supersaturated state (C=1). Hence this phase of the experiments corresponds to the growth process with transport across a diffusion boundary layer attached to the crystal. With the passage of time, the crystal size increases and the concentration gradients increase in strength. At this stage, the concentration field is accompanied by a partial loss of symmetry. Experiments of the present study revealed a marginal unsteadiness as well. This phase defines a transition process from diffusion-dominated transport of solute to the onset of convection. With the passage of time, a weak convection current is set up around the crystal. The images recorded at t=70 hours also show a greater loss of symmetry in the concentration contours and a denser spacing closer to the crystal are to be seen.

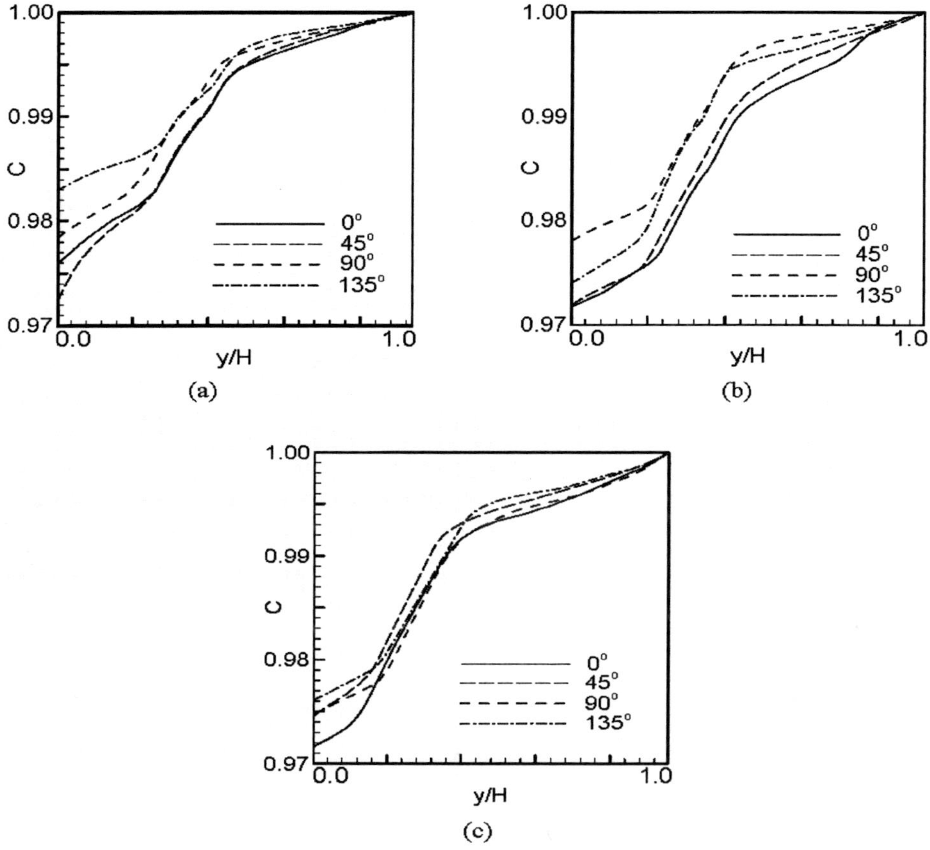

Figure 28. Width-averaged normalized concentration profile as a function of vertical coordinate for the diffusion-dominant crystal growth. Time elapsed (a) 1 hour, (b) 35 hours and (c) 70 hours.

Figure 28 shows the variation of concentration with respect to the (vertical) y-coordinate in the crystal growth chamber. The concentration referred here is an average obtained in the viewing direction of the laser beam and across the image as well. For averaging the concentration values in the x-direction, namely normal to the light beam, 11 columns along the width were selected. The concentration profiles for t=1 hour (Figure 28(a)) overlap in the bulk of the solution above the crystal for all the four view angles, thus revealing a uniform distribution of solute concentration in this region. At t=35 hours (Figure 28(b)), the differences in the profiles are quite noticeable. These differences can be attributed to the temporary unsteadiness that sets in when a transition from diffusion-dominated growth takes place towards convection. With the passage of time, the profiles become increasingly symmetric and overlap with each other. The growth now takes place in the presence of an upward rising convective plume (Figure 28(c)). The symmetry in concentration distribution indicates an almost uniform and symmetric deposition of solute onto the crystal surfaces. It is to be noticed that for the three time instants, the profiles show differences in the vicinity of the crystal (y/H=0.0). The morphology of the KDP crystal is pyramidal, as shown in Figure 29 in a photograph of a grown KDP crystal in the diffusion-dominated regime, at the end of 70 hours. The differences in the average concentration near the crystal can be traced to the non-symmetric shape of the crystal as well as large gradients at the crystal edges.

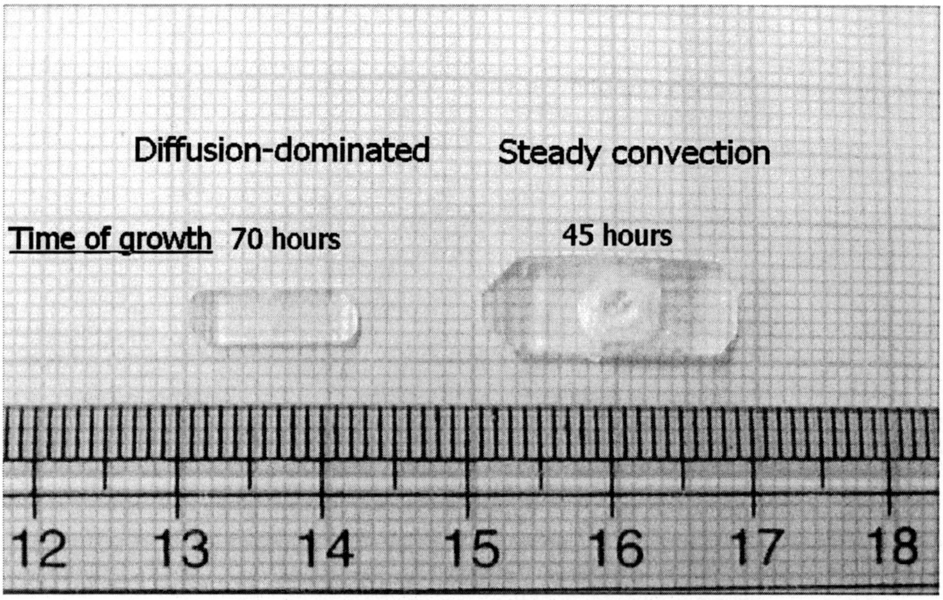

Figure 29. Photographs of as grown KDP crystals in diffusion-dominated (after 70 hours) and stable convection growth regimes (after 45 hours).

Figure 30 shows the concentration profiles reconstructed over five horizontal planes (y/H = 0.05, 0.15, 0.45, 0.60 and 0.90) above the growing crystal. The location y/H = 0.05 is just above the crystal while y/H = 0.90 represents a plane close to the top of the recorded schlieren image. The reconstructions correspond to the images presented in Figure 26. The outer circle represents the periphery of the beaker in which the crystal growth process is conducted. For a given time instant the images become progressively darker away from the crystal revealing the presence of supersaturated solution in the far field. At later times, as well as away from the crystal, the solution closer to the walls of the beaker is seen to be supersaturated, and the field is close to axisymmetric since the concentration contours here are circular. There is a significant departure from axisymmetry in the initial phase of crystal growth, as well as for longer times near the crystal itself. The second factor is explained by the fact that the crystal itself has a pyramidal structure and lacks circular symmetry. In the initial stages of growth, solutal transport is dictated by molecular diffusion from the solution to the crystal surface. Consequently, the lack of symmetry in the crystal morphology is felt on planes away from the crystal itself. At later times, a weak convection plume introduces the symmetry pattern of the beaker in the solute distribution. Hence, a degree of regularity is seen in the concentration contours, particularly on planes away from the crystal.

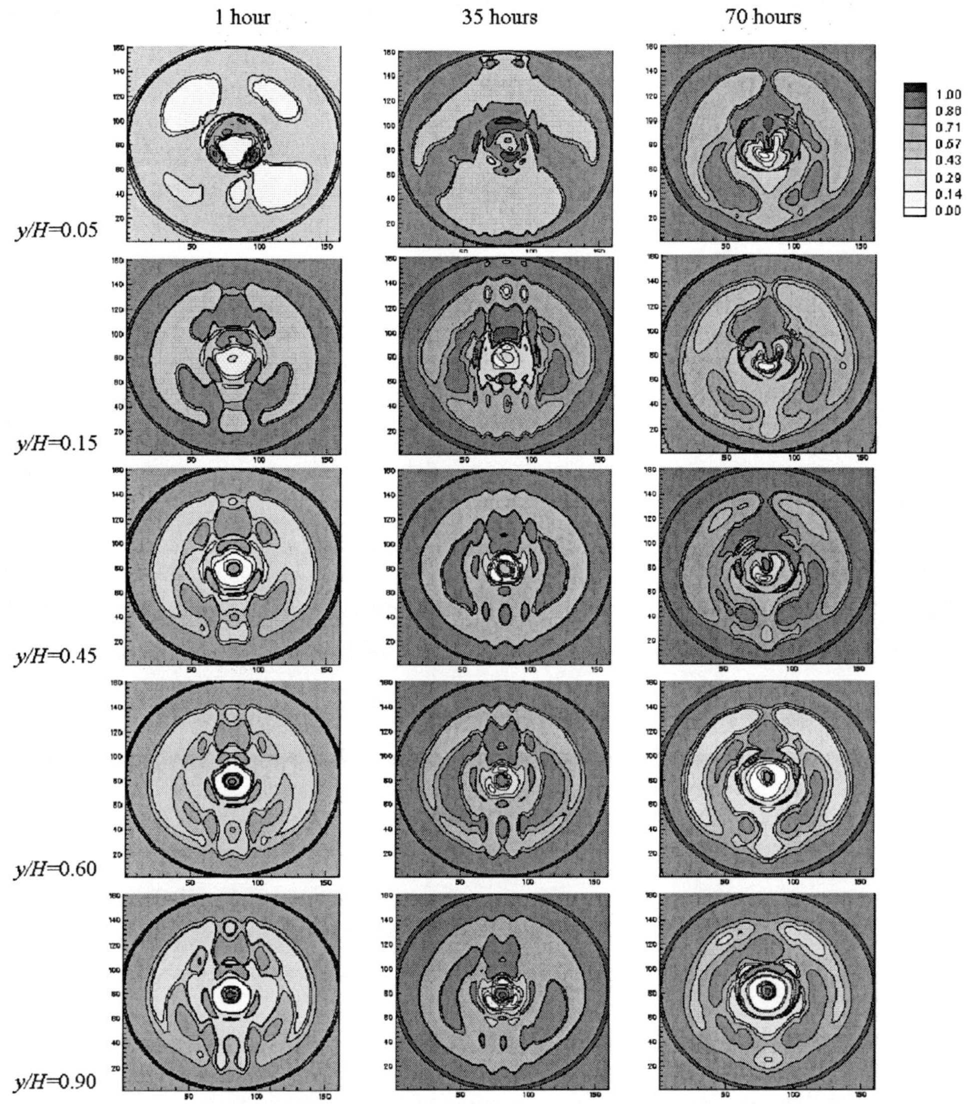

Figure 30. Reconstructed concentration profiles in the diffusion-dominated regime over five horizontal planes covering the range *y/H*=0.05-0.90, *H* being the vertical extent of the recorded schlieren image above the growing crystal.

There are several interesting features to be noticed in Figure 30.

1. At the central core of the reconstructed cross-section, the dark shade indicates the presence of a supersaturated solution. The surrounding region is bright, indicative of solution depleted of salt. Thus, the deposition of the salt on the crystal is seen to occur on the periphery of the crystal. This is a region of bright light intensity (Figure

26) and hence high concentration gradients. The relationship between high gradients and preferential growth rates of the crystal faces is brought out in Section 9.2.

2. The concentration contours near the crystal (y/H=0.05 and 0.15) show significant departure from symmetry during the initial stages of the growth process as well as during the transition phase (t=35 hours). The transition referred here is the appearance of a convection plume that is gradually set up around the crystal, resulting in the partial unsteadiness in fluid motion. The loss of symmetry in the reconstruction can be traced to the unsteadiness as well. Symmetry is partly restored on planes above y/H=0.45 at t=35 hours, indicating the influence of uniform solute concentration in the bulk of the solution. For t=70 hours, the convection plume is steady and the reconstructions are better structured closer to the crystal as well (for example, y/H=0.15).
3. The spacing of contours in the central region is smaller for the convection regime, when compared to diffusion. Thus, convection increases the concentration gradients near the crystal, increases the salt deposition rate and hence the overall increase in crystal size (Figure 30). This is possibly the most suitable transport regime for crystal growth.

9.4.2. Growth Patterns in the Convection Regime

Crystal growth from a solution in which the limit of supersaturation has been reached is now considered. The density differences (ramp rate of 0.05°C per hour) as well as the crystal size (6 mm) are jointly large enough to sustain fluid motion in the beaker from the start. Experiments have been conducted under conditions of a steady plume rising from the crystal. Results for 30 hours of growth have been presented. For times greater than 30 hours, the plume became quite unsteady. Consequently, the projection data of individual view angles became uncorrelated.

Figure 31 shows the convective field in the form of schlieren images. A well-defined plume rising from the top surface of the crystal is seen at both times of t=2 and 30 hours for four view angles. At t=2 hours, the plume was temporally stable , though a swaying movement was seen from one projection angle to the other. The swaying motion was due to the disturbance of the flow field caused by the turning of the growth chamber for recording the projection data. The tilt in the plume at selected angles for small time is also related to the non-symmetric shape of the initial crystal itself. At a later time (t=30 hours), the gradients are large enough to give rise to a stable and symmetric convective field. This is accompanied by uniform deposition of the solute on the crystal surfaces. As discussed in Section 9.4.1, this regime yields the highest quality of the crystal in terms of symmetry of the faces and its transparency, along with the fastest growth rate.

Figure 32 shows concentration contours around the growing crystal derived from the schlieren images of Figure 31. The physical extent of the convective plume in the vertical direction can be estimated by the deformation of the concentration contours. Densely placed contours in the crystal vicinity are due to high concentration gradients. The contours at t=30 hours are seen to spread into the growth chamber, indicating that solute from the bulk of the solution itself is deposited over the crystal. The appearance of double peaks in the contour shape at the end of 30 hours confirms that the highest gradients and hence the deposition rates occur towards the edges of the crystal.

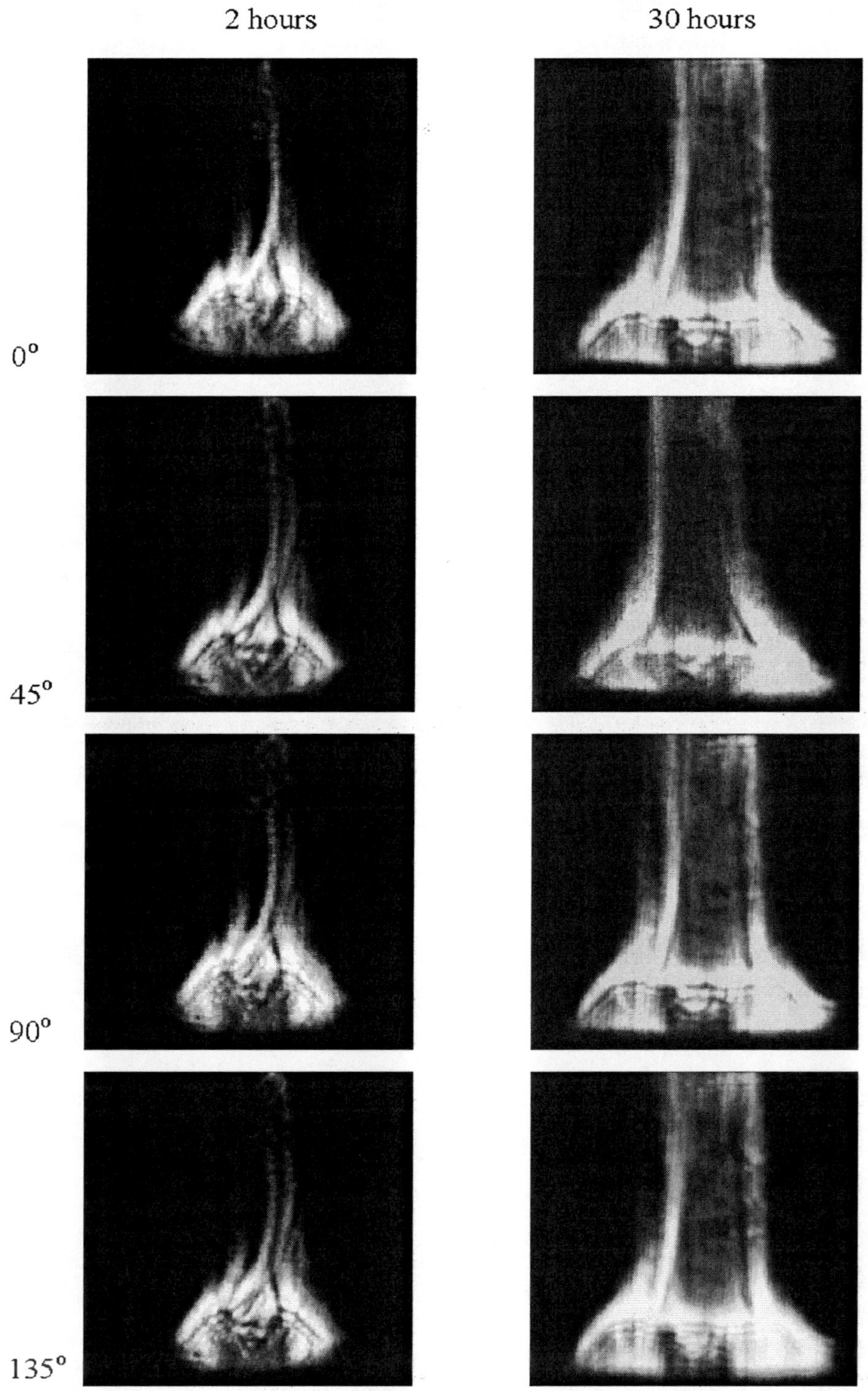

Figure 31. Schlieren images of the convective field around crystal growing from its aqueous solution as recorded from four different view angles at two different times instants (2 and 30 hours).

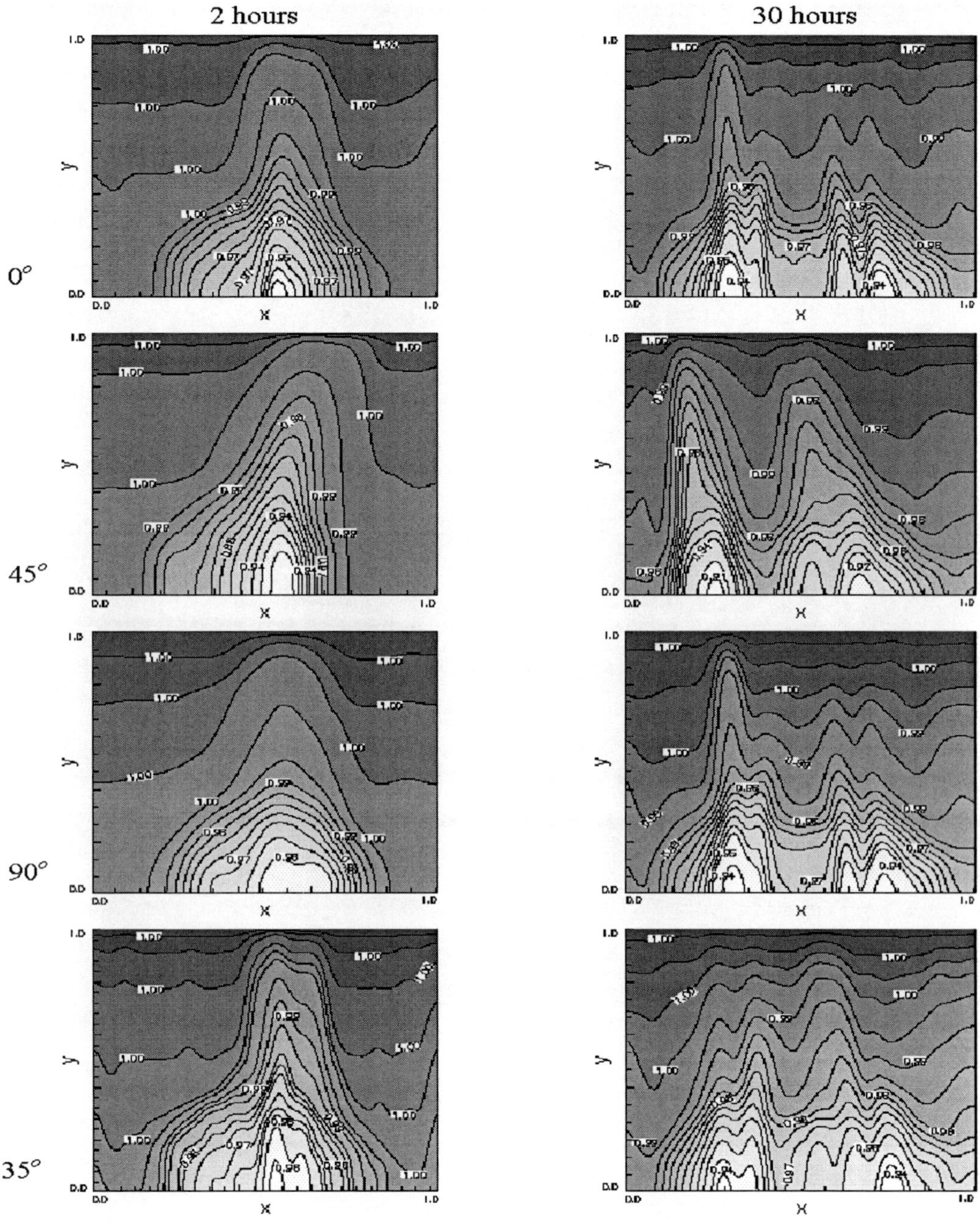

Figure 32. Concentration contours around a growing crystal in the presence of well-defined convective plume for the four view angles at two different time instants.

Figure 33 shows the distribution of width-averaged concentration with respect to the vertical coordinate at two time instants, namely 2 and 30 hours. The concentration data presented in the figure is integrated along the viewing direction by the light beam, and once again across the width of the image. The differences among the individual profiles for each of the view angles are a measure of loss of symmetry in the concentration distribution above the crystal. At t=2 hours, the differences are large in the bulk of the solution ($y/H \geq 0.55$). The difference diminishes at locations closer to the crystal (*y/H*=0.0), Figure 33(a). With the passage of time, the convective plume stabilizes in terms of symmetry. The average concentration data corresponding to *t*=30 hours for the four view angles reveals a good degree of symmetry in the entire field of view above the crystal (Figure 33(b)).

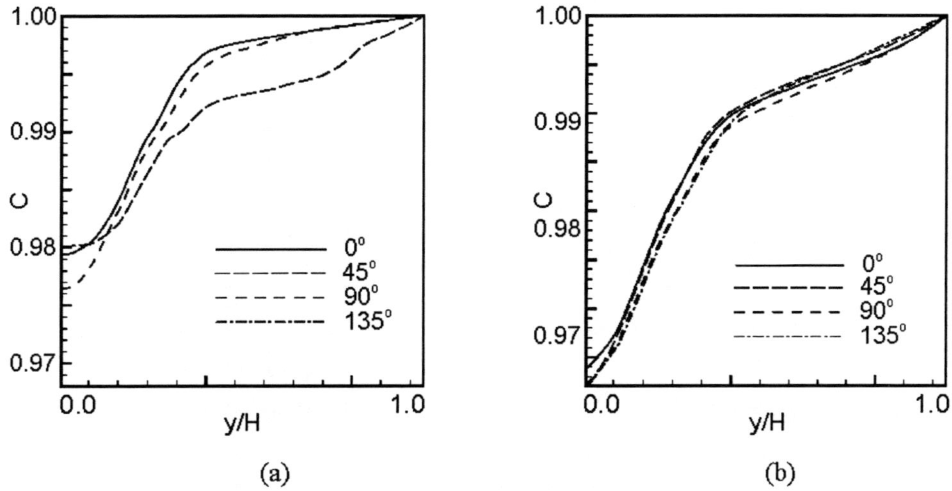

Figure 33. Width-averaged normalized concentration profile as a function of the vertical coordinate around a growing crystal in the stable convection regime after a period of (a) 2 hours, and (b) 30 hours.

Figure 34 shows reconstructed concentration profiles at selected horizontal planes above the growing crystal for the two time instants considered in Figure 31. The following observations can be recorded on the basis of the reconstruction.

1. At *t*=2 hours, the reconstructed concentration field reveals symmetry for the plane close to the crystal (*y/H*=0.05) both around the crystal and in the bulk of the solution. The presence of concentric rings indicates uniform distribution of concentration gradients on all the faces of the growing crystal. This feature is seen in Figure 31 as well, where the bright intensity region uniformly envelops the crystal faces. For planes away from the crystal (*y/H*≥0.30), the deformation of the concentric rings in the vicinity of the crystal is mainly due to the swaying motion of the convective plume introduced by turning the growth chamber. This movement is also to be seen in the schlieren images of Figure 31. On the other hand, the bulk of the solution closer to the beaker shows uniform and symmetric distribution of concentration. The appearance of only a few contours in the solution shows low concentration gradients here in the initial stages of growth.
2. For all planes (*t*=2 hours), the presence of a bright central core of the reconstructed cross-section is indicative of the solution depleted of salt. The spatial extent of the bright region at the center of the beaker is small, in agreement with the thin vertically rising buoyant plume above the growing crystal, Figure 31.

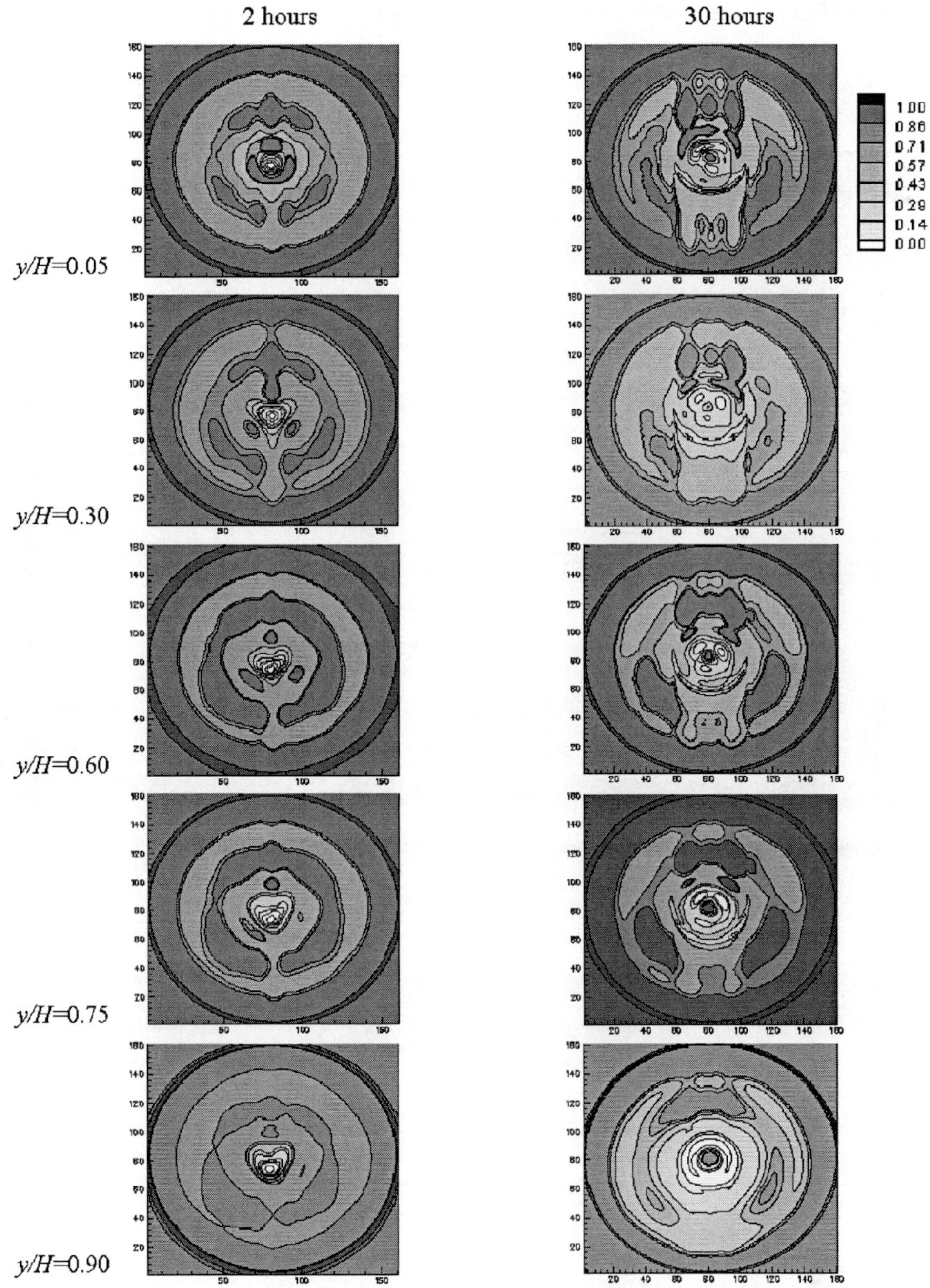

Figure 34. Reconstructed concentration profiles over five horizontal planes (y/H=0.05-0.90) above the crystal growing in the presence of a stable convection plume. Time instants of 2 and 30 hours of experimental run time are presented.

3. At later times (t=30 hours), the concentration field at planes near the crystal ($y/H\leq 0.30$) is not axisymmetric though the bulk of the solution near the boundaries of the growth chamber shows the presence of concentric rings. The loss of symmetry at the center of the reconstructed plane can be attributed to the role played by the

crystal geometry in deciding the overall orientation of convection pattern. The effect of crystal shape is negligible when the size is small (for example, at t=2 hours) but is significant when the crystal grows in size with a well-defined morphology (Figure 29). In addition, densely spaced contours in the core and the bulk reflect the presence of high concentration gradients in the growth chamber.

4. For planes located away from the crystal ($y/H \geq 0.60$), the effect of crystal geometry on the convective field is diminished. The presence of a dark central core (corresponding to a higher level of supersaturation) surrounded by a brighter region shows that high concentration gradients prevail on the crystal sides in comparison to its upper face.

10. CONCLUSIONS

The characterization of convection patterns around a crystal growing from its aqueous solution is the subject of the present investigation. The study has been conducted using three laser measurement techniques that rely on the variation of refractive index with density, and hence the solutal concentration. These techniques have been validated in a thermal buoyancy experiment in a rectangular cavity. Subsequently, they have been used to study convection patterns around the growing crystal for varying process parameters. While images contain line-of-sight averaged information in a vertical plane, the possibility of reconstructing concentration contours on horizontal planes has been studied *via* tomography. Experiments reported in the study show that convection does influence crystal growth rate as well as its quality. The laser measurement techniques succeed in capturing the solutal concentration field (or its derivatives) around the crystal. Hence they can be used for online process control. Specific conclusions arrived at are given in the following sections.

10.1. Comparison of Optical Techniques

Three refractive index-based optical techniques, namely interferometry, schlieren and shadowgraph have been applied to image the convective flow field in a differentially heated rectangular cavity. Air and water have been considered as working fluids. Experiments have been conducted over a wide range of temperature differences. The following conclusions have been arrived at in the present work.

1. In low temperature gradient experiments, all the three techniques correlate well with one another. Interferograms are limited by few fringes in air, and too many fringes in water. The shadowgraph image does not show sufficient contrast for analysis. In this respect, the schlieren technique is most amenable to data reduction.
2. In high gradient experiments, both schlieren and shadowgraph yield clear images. The interferograms are however corrupted by refraction errors. Schlieren and shadowgraph track the temporal response of the fluid medium in the form of the light intensity variation.

3. In high Rayleigh number experiments with water, the flow field is turbulent. Shadowgraph images are seen to be meaningful, as against interferograms and schlieren. They reveal a considerable amount of physical information, including boundary-layers, plumes and time scales.

10.2. Optical Techniques Applied to Crystal Growth

A direct comparison of interferometry, schlieren and shadowgraph in the context of growth of optical crystals from an aqueous solution has been conducted. The three techniques reveal similar trends in terms of the spatial distribution of concentration gradients in the vicinity of the growing crystal. The initial dissolution and unsteadiness, a stable and symmetric growth regime and the final stratification of the solution are brought out. The free convection crystal growth experiment is thus limited in the long run by the localization of high concentration gradients around the growing crystal. The stratification here is stable and suppresses fluid motion. This limits the size of the crystal that can be grown. Based on the ease of instrumentation, image clarity and simplicity of data analysis, the laser schlieren technique emerges as better suited for process control, as compared to interferometry and shadowgraph.

10.3. Influence of Process Parameters on Crystal Growth

The dependence of the convection patterns, concentration gradients and the crystal growth rate on ramp rate, rotation and crystal size is examined. Laser schlieren technique is employed as the measurement tool for determining the concentration distribution in the solution. Major conclusions to emerge from the study are:

1. The stable growth regime of the crystal at 0 rpm comprises of symmetric buoyancy-generated convection plumes. When rotation is imparted to the crystal, the stable growth regime reveals a helical flow path.
2. Buoyancy-driven convection intensifies with increasing size of the crystal as well as the ramp rate. It is ultimately limited by salt depletion and solute stratification in the growth chamber that bring the fluid particles to rest. Crystal rotation diminishes the concentration gradients, equalizes them over all faces, and improves the symmetry of the concentration distribution in the solution. This leads to a crystal of better quality. The rpm of the crystal required to improve crystal quality depends on the cooling rate of the solution.
3. Over the range of parameters studied, the large-scale convection pattern is buoyancy-driven, even when the crystal is rotated.
4. Growth rates of the crystal correlate with the magnitude of the concentration gradients over its individual faces. The gradients in turn respond to the convective field around the crystal.
5. Experiments with crystals of various sizes immersed in an aqueous solution show that the convection transients are short-lived. Thus, the growth process is quasi-steady and can be controlled by adjusting the ramp rate and crystal rotation.

10.4. Three Dimensional Reconstruction

Concentration distribution around a growing KDP on selected horizontal planes is reported. Partial images recorded at various view angles have been analyzed using the convolution back projection algorithm. A suitable extrapolation scheme has been employed to generate information about the entire measurement volume using the partial projection data. The performance of the extrapolation scheme has been assessed by reconstructing the numerically simulated three-dimensional temperature field in an axisymmetric differentially heated fluid layer. Steady crystal growth experiments have been conducted in two regimes that respectively are dominated by diffusion and buoyancy. The following conclusions emerge from the present study:

1. The extrapolation technique used in the present work was seen to perform adequately with simulated data. In particular, no new artifacts were seen even when the original data set was as small as 30%.
2. The concentration field in the vicinity of a crystal growing in a diffusion-dominated regime showed symmetry along the centerline of the beaker indicating a uniform deposition of solute on the crystal surfaces over a long period of time. The field in the bulk of the solution was entirely axisymmetric in the initial stages of the growth process.
3. The transition from diffusion-dominated growth to the onset of convection induced temporary unsteadiness in the flow field that was accompanied by a breakdown of the symmetry of the concentration field.
4. With an increase in the crystal size, the convection current was seen to increase in strength and thus determine the overall transport of solute from the bulk of the solution to the crystal surfaces. The geometry of the grown crystal fixed the orientation of the rising plume. Nearly symmetric distribution of concentration was realized in the stable growth regime in regions away from the crystal.

11. Future Work

In the context of measurements, we propose to analyze shadowgraph images of convection patterns, collect tomographic projections by synchronizing the crystal rotation with the frame rate of the camera, and implement grey-scale and color-filters for schlieren measurements. The suitability of iterative tomographic algorithms will be explored. The interaction of crystal rotation with buoyancy will be studied over a wider range of parameters. A closer connection between the facets of the crystal and convection patterns will be sought to be established.

Acknowledgements

The importance of fluid convection in many crystal growth processes was originally pointed out to the authors by Dr. Vishwanath Prasad, Dean of Engineering, Florida State

University, Florida. KM thanks Dr. V.K. Wadhawan, CAT Indore (India) for introducing him to crystal growth from an aqueous solution. AS is grateful to Department of Atomic Energy, Mumbai for awarding him the K.S. Krishnan Fellowship for the year 2004. The authors held several useful discussions on the aqueous solution technique with Mr. Sunil Verma, CAT Indore (India). Financial support of Department of Science and Technology, New Delhi during 2000-2003 is thankfully acknowledged.

NOMENCLATURE

A	Archimedes number, Ra/Re^2
C	Solute concentration, kmol of salt/kg of solution
ΔC_ε	Concentration difference between two successive fringes
ΔC	Concentration difference between the saturated and supersaturated solution
D	Distance of the screen from the optical window (also diameter of the growth chamber), m
E_1	Absolute maximum temperature difference
E_2	RMS error
f	Focal length of the de-collimating mirror, m
g	Acceleration due to gravity, m/s^2
H	Height of the fluid layer (rectangular cavity), m (also, image size as seen through the optical window)
I	Intensity distribution (digitized)
L	Distance traversed by the laser beam through the test cell (also, crystal thickness in the vertical direction), m
N	Molar concentration of the solution, moles per 100 gm of KDP solution (also, total number of grid points on the reconstructed plane for tomography reconstruction)
Nu	Local Nusselt number defined as wall heat flux×cavity height/fluid conductivity
n	Refractive index of the fluid
n_a	Refractive index of the ambient
Pr	Prandtl number of the fluid, ν/α
R	Spatial frequency in the CBP algorithm
Ra	Rayleigh number (thermal= $g\,\beta\,\Delta T\,H^3/\,\nu\alpha$)
Re	Rotational Reynolds number, $\omega L^2/\,\nu$
r	radial coordinate
s	Perpendicular distance of the data ray from the center of the object
t	Time, hr
T	Temperature, K
T_o	Reference temperature, K
ΔT_ε	Temperature difference between successive fringes, K
$W(R)$	Filter function
x,y,z	Cartesian coordinates, with z parallel to the direction of propagation of

light

Greek Symbols

α	Diffusivity (mass/thermal), m^2/s (also polarizability of the KDP solution)
β	Coefficient of volume expansion of the fluid, K^{-1}
λ	Wavelength of the laser, nm
δ	Angular deflection of laser beam
ϕ	Polar angle in the CBP algorithm
ν	Kinematic viscosity of the solution, m^2/s
ρ	Density of the solution, kg/m^3
ρ_o	Reference value of density, kg/m^3
θ	Position of the source-detector line with respect to a fixed axis
ω	Crystal rotation speed, rpm

Subscripts

a,o	Reference value
c	Cold
h	Hot
orig	Original (temperature)
recon	Reconstructed (temperature)

REFERENCES

[1] W.R. Wilcox, "Transport phenomena in crystal growth from solution", *Prog. Crystal Growth and Charact.* 26, 153-194 (1993).

[2] Gebhart B., Jaluria Y., Mahajan R.L., and Sammakia B., *Buoyancy-Induced Flows and Transport,* Hemisphere Publishing Corporation, New York (1988).

[3] R.W. Lewis, R.E. Teets, J.A. Sell, and T.A. Seder, "Temperature measurements in a laser-heated gas by quantitative shadowgraphy", *Applied Optics* 26(17), 3695-3704 (1987).

[4] S. Kosugi, K. Maeno, and H. Honma, "Measurement of gas temperature profile in discharge region of excimer laser with laser schlieren method", *Japanese Journal of Applied Physics*, 32(3), 4980-4986 (1993).

[5] K. Muralidhar, V.B. Patil, and R. Kashyap, "Interferometric study of transient convection in a square cavity", *Journal of Flow Visualization and Image Processing,* 2(4), 321-333 (1995).

[6] M.S. Chandrashekhara, D.D. Squires, M.C. Wilder, and L.W. Carr, "A phase-locked high-speed real-time interferometry system for large amplitude unsteady flows", *Experiments in Fluids* 20, 61-67 (1995).

[7] S. Zhong and L.C. Squires, "An interferometric study of organized structures in wakes of circular cylinders", *Experiments in Fluids* 20, 91-99 (1995).

[8] O. Dupont, J.L. Dewandel, and J.C. Legros, "Use of electronic speckle pattern interferometry for temperature distribution measurements through liquids", *Optics Letters* 20(17), (1995).

[9] W. Schopf, J.C. Patterson, and A.M.H. Brooker, "Evaluation of the shadowgraph method for the convective flow in a side-heated cavity", *Experiments in Fluids*, 21, 331-340 (1996).

[10] G. Tanda, and F. Devia, "Application of a schlieren technique to heat transfer measurements in free convection", *Experiments in Fluids*, 24, 285-290 (1998).

[11] L. Howle, R.P. Behringer, and J. Georgiadis, "Visualization of convective fluid flow in a porous medium", *Nature* 362, 230-232 (1993).

[12] A.K. Agrawal, N.K. Butak, S.R. Gollahalli, and D. Griffin, "Three-dimensional rainbow schlieren tomography of a temperature field in gas flows", *Applied Optics* 37(3), 485-497 (1998).

[13] W. Dezhong and Z. Tiange, "The measurement of 3-D asymmetric temperature field by using real time laser interferometric tomography", *Optics and Lasers in Engineering* 36, 289-297 (2001).

[14] R. Snyder and L. Hesselink, "Measurement of mixing fluid flows with optical tomography", *Optics Letters* 13(2), 87-89 (1988).

[15] D.W. Watt and C.M. Vest, "Turbulent flow visualization by interferometric integral imaging and computed tomography", *Experiments in Fluids* 8, 301-311 (1990).

[16] Y.C. Michael and K.T. Yang, "Three-dimensional Mach-Zehnder interferometric tomography of the Rayleigh-Benard problem", *ASME J. Heat Transfer* 114, 622-629 (1992).

[17] L. McMackin and R.J. Hugo, "High speed optical tomography system for imaging dynamic transparent media", *Optics Express* 1(11), 302-311 (1997).

[18] H.S. Ko and K.D. Kihm, "An extended algebraic reconstruction technique (ART) for density-gradient projections: laser speckle photographic tomography", *Experiments in Fluids* 27, 542-550 (1999).

[19] D. Mishra, K. Muralidhar, and P. Munshi, "Study of Rayleigh-Benard convection at intermediate Rayleigh numbers", *Fluid Dynamics Research*, 25(5), 231-255 (1999).

[20] D. Mishra, J.P. Longtin, R.P. Singh, and V. Prasad, "Performance evaluation of iterative tomography algorithms for incomplete projection data", *Applied Optics* 43(7), 1-11 (2004).

[21] P.S. Chen, P.J. Shlichta, W.R. Wicox, and R.A. Lefever, "Convection phenomena during the growth of sodium chlorate crystals from solution", *Journal of Crystal Growth* 47, 43-60 (1979).

[22] P.S. Chen, W.R. Wicox, and P.J. Shlichta, "Free convection about a rectangular prismatic crystal growing from a solution", *Int. J. Heat Mass Transfer* 22, 1669-1679 (1979).

[23] P.J. Shlichta, "Feasibility of mapping solution properties during the growth of protein crystals", *Journal of Crystal Growth* 76, 656-662 (1986).

[24] F. Bedarida, "Developments of holographic interferometryapplied to crystal growth from solution", *Journal of Crystal Growth* 79, 43-49 (1986).

[25] V.I. Bredikhin, V.P. Ershov, V.V. Korolikhin, V.N. Lizyakina, S.Yu. Potapenko, and N.V. Khlyunev, "Mass transfer processes in KDP crystal growth from solutions", *Journal of Crystal Growth* 84, 509-514 (1987).

[26] K. Onuma, K. Tsukamoto, and I. Sunagawa, "Role of buoyancy convection in aqueous solution growth; A case study of $Ba(NO_3)_2$ crystal", *Journal of Crystal Growth* 89, 177-188 (1988).

[27] K. Onuma, K. Tsukamoto, and I. Sunagawa, "Effect of buoyancy driven convection upon the surface microtopographs of $Ba(NO_3)_2$ and CdI_2 crystals", *Journal of Crystal Growth* 98, 384-390 (1989).

[28] K. Onuma, K. Tsukamoto, and I. Sunagawa, "Measurements of surface supersaturations around a growing K-Alum crystal in aqueous solution", *Journal of Crystal Growth* 98, 377-383 (1989).

[29] F. Bedarida, E. Piano, and C. Pontiggia, "Measurements of concentration profiles by diffraction techniques", *Journal of Crystal Growth* 106, 217-220 (1990).

[30] M. Mantani, M. Sugiyama, and T. Ogawa, "Electronic measurement of concentration gradient around a crystal growing from a solution by using Mach-Zehnder interferometer", *Journal of Crystal Growth* 114, 71-76 (1991).

[31] K. Onuma, K. Tsukamoto, and S. Nakadate, "Application of real time phase shift interferometer to the measurement of concentration field", *Journal of Crystal Growth* 129, 706-718 (1993).

[32] I.Sunagawa, K. Tsukamoto, K. Maiwa, and K. Onuma, "Growth and perfection of crystals from aqueous solution: Case studies on barium nitrate and K-Alum", *Prog. Crystal Growth and Charact.*, 30, 153-190 (1995).

[33] X. Yu, Y. Liu, X. Yu, and Y. Sun, "Some new optical measurement techniques for the study of crystal growth and electrode processes", *Optics and Lasers in Engineering* 25, 191-204 (1996).

[34] S. Kleine, W.J.P. van Enckevort, and J. Derix, "A dark field-type schlieren microscope for quantitative, in situ mapping of solute concentration profiles around growing crystals", *Journal of Crystal Growth* 179, 240-248 (1997).

[35] M.J. Krasinski, "Evolution of diffusion fields around KDP crystals growing in gel", *Cryst. Res. Technol.* 34(5-6), 647-653 (1999).

[36] V.I. Bredikhin, G.L. Galushkina, and S.P. Kuznetsov, "Schlieren technique to in situ monitor-rapidly growing KDP crystal surface", *Journal of Crystal Growth* 219, 83-90 (2000).

[37] T. Inoue, K. Mori, Y. Kageyama, and H. Mori, "Various growth shapes of $Na_2S_2O_3.5H_2O$ crystals", *Cryst. Res. Technol.* 35(5), 587-593 (2000).

[38] E. Piano, G.A. Dall'Aglio, S. Crivello, R. Chittofrati, and F. Puppo, "New optical techniques for crystal growth from fluids", *Material Chemistry and Physics* 66, 266-269 (2000).

[39] E. Piano, G.A. Dall'Aglio, R. Chittofrati, S. Crivello, and F. Puppo, "A non-destructive interferometric technique for analysis of crystal growth and fluid dynamics", *Ann. Chim. Sci. Mat.* 26, 23-28 (2001).

[40] L. Duan, and J.Z. Shu, "The convection during $NaClO_3$ crystal growth observed by the phase shift interferometer", *Journal of Crystal Growth* 223, 181-188 (2001).

[41] N.A. Booth, A.A. Chernov, and P.G. Vekilov, "Characterestic lengthscales of step bunching in KDP crystal growth: in situ differential phase-shifting interferometry study", *Journal of Crystal Growth* 237-239, 1818-1824 (2002).

[42] S. Maruyama, K. Ohno, A. Komiya, and S. Sakai, "Description of the adhesive crystal growth under normal and micro-gravity conditions employing experimental and numerical approaches", *Journal of Crystal Growth* 245, 278-288 (2002).

[43] V.I. Bredikhin, G.L. Galushkina, A.A. Kulagin, S.P. Kuznetsov, and O.A. Malshakova, "Competing growth centers and step bunching in KDP crystal growth from solutions", *Journal of Crystal Growth* 259, 309-320 (2003).

[44] A. Srivastava, K. Muralidhar, and P.K. Panigrahi, "Comparison of interferometry, schlieren and shadowgraph for visualizing convection around a KDP crystal", *Journal of Crystal Growth* 267(1), 348-361 (2004).

[45] Atul Srivastava, K. Muralidhar, and P.K. Panigrahi, "A schlieren study of the effect of ramp rate and rotation on convection around a crystal growing from an aqueous solution", *Journal of Crystal Growth* 274(1-2), 191-208 (2005).

[46] S. Miyashita, H. Komatsu, Y. Suzuki, T. Nakada, "Observation of the concentration distribution around a growing lysozyme crystal", *Journal of Crystal Growth* 141, 419-424 (1994).

[47] P.G. Vekilov and F. Rosenberger, "Protein crystal growth under forced solution flow: experimental set up and general response of lysozyme", *Journal of Crystal Growth* 186, 251-261 (1998).

[48] W.B. Hou, A.B. Kudryavtsev, T.L. Bray, L.J. DeLucas, and W.W. Wilson, "Real time evolution of concentration distribution around tetragonal lysozyme crystal: case study in gel and free solution", *Journal of Crystal Growth* 232, 265-272 (2001).

[49] D. Yin, Y. Inatomi, and K. Kuribayashi, "Study of lysozyme crystal growth under a strong magnetic field using a Mach-Zehnder interferometer", *Journal of Crystal Growth* 226, 534-542 (2001).

[50] G. Sazaki, Y. Azuma, S. Miyashita, N. Usami, T. Ujihara, K. Fuziwara, Y. Murakami, K. Nakajima, "*In-situ* monitoring system of the position and temperature at the crystal-solution interface", *Journal of Crystal Growth* 236, 125-131 (2002).

[51] M. Kawaji, O. Gamache, D.H. Hwang, N. Ichikawa, J.P. Viola, and J. Sygusch, "Investigation of Marangoni and natural convection during protein crystal growth", *Journal of Crystal Growth* 258, 420-430 (2003).

[52] N. Ramachandran and F.W. Leslie, "Using magnetic field to control convection during protein crystallization-analysis and validation studies", *Journal of Crystal Growth* 274, 297-306 (2005).

[53] L. Gatti, F. Solitro, F. Bedarida, P. Boccacci, G.A. Dall'aglio, and L. Zefiro, "Three-dimensional measurements of concentration fields in crystal growth by multidirectional holographic interferometry," in *Laser Interferometry: Quantitative analysis of Interferograms*, Proc. SPIE 1162, 126-131 (1989).

[54] D.W. Sweeney and C.M. Vest, "Reconstruction of three-dimensional refractive index fields from multidirectional interferometric data," *Applied Optics* 12(11), 2649-2664 (1973).

[55] A.G. Notcovich, I. Braslavsky, and S.G. Lipson, "Imaging fields around growing crystals," *Journal of Crystal Growth* 198/199, 10-16 (1999).

[56] A Srivastava, K. Muralidhar, and P.K. Panigrahi, "Reconstruction of the concentration field around a growing KDP crystal using schlieren tomography", (manuscript in preparation).

[57] N.P. Zaitseva, L.N. Rashkovich, and S.V. Bogatyreva, "Stability of KH_2PO_4 and $K(H,D)_2PO_4$ solutions at fast crystal growth rates", *Journal of Crystal Growth* 148, 276-282 (1995).

[58] R.J. Goldstein, ed., *Fluid Mechanics Measurements,* second edition, (Taylor and Francis, New York, 1996).

[59] G.S. Settles, *Schlieren and shadowgraph techniques: Visualizing phenomena in transport media* (Springer-Verlag publications, New York, 2001).

[60] D. Mishra, K. Muralidhar, and P. Munshi, Performance Evaluation of Fringe Thinning Algorithms for Interferometric Tomography, *Optics and Lasers in Engineering*, Vol. 30, pp 229-249, (1998).

[61] G.T. Herman, *Image Reconstruction from Projections* (Academic, New York, 1980).

[62] F. Natterer, *The Mathematics of Computerized Tomography* (John Wiley, New York, 1986).

[63] F. Mayinger, ed., *Optical Measurements: Techniques and Applications* (Springer-Verlag, Berlin, Germany, 1994).

[64] K. Muralidhar, Temperature Field Measurement in Buoyancy-driven Flows using Interferometric Tomography, *Annual Review of Heat Transfer*, Vol. 12, pp. 265-376, 2001.

[65] P. Munshi, "Error analysis of tomographic filters. I: Theory," *NDT & E International* 25(4/5), 191-194 (1992); Part II: Results, *NDT & E International* 26(5), 235-240 (1993).

[66] M.G. Velarde and C. Normand, "Convection," *Scientific American* 243(1), 79-94 (1980).

[67] S.S. Leong, "Numerical study of Rayleigh-Benard convection in a cylinder," *Numerical Heat Transfer* (A) 41, 673-683 (2002).

[68] S. Punjabi, K. Muralidhar, and P.K. Panigrahi, Buoyancy-driven Convection in Superposed Fluid Layers in an Octagonal Cavity, *International Journal of Thermal Sciences*, Vol. 43(9), pp. 849-864 (2004).

[69] A.Srivastava, K. Muralidhar, and P.K. Panigrahi, Interferometric study of buoyancy-driven convection in a circular fluid layer, *Heat and Mass Transfer*, Vol. 41(4), pp. 353-359 (February 2005).

[70] Srivastava, A. Phukan, P.K. Panigrahi and K. Muralidhar, Imaging of a Convective Field in a Rectangular Cavity using Interferometry, Schlieren and Shadowgraph, *Optics and Lasers in Engineering*, Vol. 42(4), pp. 469-485 (2004).

[71] J.P. Vanyo, *Rotating Fluids in Engineering and Science,* Dover Publications, New York, (1993).

In: New Topics in Crystal Growth Research
Editor: George V. Karas, pp. 85-124
ISBN 1-59454-837-4

Chapter 2

HYDRO/SOLVOTHERMAL ROUTE TO GROWTH OF NANOCRYSTALS

Yuanfang Liu [1, 2,*]
[1]Department of Chemistry, Oklahoma State University, Stillwater, OK, 74078, USA
[2]Department of Chemistry, University of Science and Technology of China, Hefei, Anhui, 230026, PR China

ABSTRACT

The hydro/solvothermal method of growing crystals has become widely accepted in the last few decades. The crystals of a variety of compounds and elements, such as simple oxides, complex oxides, chalcogenides, chalcohalides, metal elements, and non-metal elements, have been grown from hydro/solvothermal process. A few literatures and books have given reviews on the hydrothermal crystal growth of the above compounds and elements, mainly on their bulk crystal growth. In this account, we will focus on the nanocrystal growth of different types of compounds and elements under hydro/solvothermal condition. The nanostructures mainly include nanorods, nanotubes, nanoneedles, nanowires, nanospheres and nano- hollow spheres, nano-flakes, and nanodots. The influence factors, such as starting materials, solvents, pH, temperature, reaction time, on the growth of nanocrystals and on the morphology and size control of nanocrystals will be discussed since many physical and chemical properties of nanomaterials are depended on the shape and size to a large extent.

INTRODUCTION

Hydrothermal/solvothermal procedures based on the process in a sealed vessel such as an autoclave and a bomb have many advantages over other high temperature solution reactions. For example, it can avoid the release of low-boiling point solvents into environment during reaction process; the resultant particles can crystallize very well under the antogenous

* Corresponding author: Email: liuyf1588@yahoo.com

pressure. Byrappa and Yoshimura [1] described the principles and applications of hydrothermal technology, the history of hydrothermal technology, apparatus of hydrothermal techniques, physical chemistry of hydrothermal crystal growth, and hydrothermal growth of some selected crystals mainly referring to bulk crystals. They listed many valuable books and reviews in the field of hydrothermal research ranging from 1971 to 1999. Ram Seshadri et al reviewed the reports of solvethermal and hydrothermal procedures for the preparation of isolated and capped nanoparticles of some important oxides and chalcogenides materials and discussed the autogenous pressures in the solvethermal reactions [2, 3].

In this review, we limit to the preparation of nanostructures by using of hydrothermal/solvothermal techniques. These nanostructures include elements, oxides and hydroxides, chalcogenides, ternary compound, hybrid composites, III-V semiconductors. We have attempted to focus on the following points: (1) Only the preparing procedures without using high toxic volatile solution will be discussed; (2) for the materials which were studied extensively, we selected those with good properties, well-dispersion, and size range less than 20nm; (3) for the materials with special structure, we still discuss their preparation and the impact factors on the preparation without considering their size, dispersion, and properties.

2. Hydrothermal/Solvothermal Preparation of Nanostructures of Elements

2.1. Introduction

The element nanostructures prepared through hydrothermal/solvothermal technique cover a variety of substances. They include metal element nanostructures, such as Ag and Cu, Sb, Bi, In, Fe, Ni, and Ru, non-metal element nanostructures, such as Se, Te, and C. Metal nanoscale particles have been attracting much attention for their intriguing chemical and physical properties and potential applications in optics, electronics, photoelectronics, catalysis, sensing, clinical diagnostics, surface-enhances Raman scattering, information storage, energy conversion/storage, and biological labeling [4-6]. The non-metal elements Se and Te are semiconductors; so far there are a few reports on their nanostructures. Non-metal element carbon, especially carbon nanotubes have attracted much attention due to their unique properties and potential applications in numerous areas such as electronic and mechanical devices [7], hydrogen storage [8], field emission tips [9], and nanotweezers [10]. Metal and nonmetal element nanostructures including nanodots, nanoparticles, nanowires, nanotubes, nanorods, nanofibers, nanosheets, nano-hollow spheres, nanobelts, were developed through hydrothermal and solvothermal method. The starting materials, reaction temperature, reaction time and results are listed in Table 1-1 and Table 1-2.

Table 1-1. The Synthesis Conditions and Results of Element Nanostructures

Elements	The type of nanostructures	Size (nm)	Starting materials & temperature (℃) & time	Properties	References
Ag	Nanodots	4	Silver octanethiolate + H2O; 180, 12h	UV-Vis absorption 423nm	[11-12]
	nanodots	4	silver dodecanethiolate+H_2O;180, 12h		
	Chainlike, dendritic	20-100×5000	$AgNO_3$+PVP+EG; 180, 3h $AgNO_3$+PVP+EG; 180, 4h, 8h, 12h	/	
Cu	nanowires	85×10000–90000	Cu^{II}-glycerol+HPO_3^{2+} +SDBS, 120, 20h	UV-Vis absorption 590nm	[13]
Sb	Nanotubes	50-80×400-800	$SbCl_3$+NaOH+Zn+H_2O, 120, 48h	/	[14-15]
	Nanowires, tube-groove-like nano-dentritic	5-15×5000 70×4000	$SbCl_3$+Zn+ethanol, 180, 12h $SbCl_3$+Zn+methanol, 160, 12h $SbCl_3$+Zn+EG, 140, 12h	/	
Bi	nanowires	25×2500	$Bi(NO_3)_3$+$H_2NCH_2CH_2NH_2$, 160, 6h	/	[16-17]
	nanotubes	2×6×500	Bi_2O_3+EG, 200, 12h	/	
In	Nanotubes,	60×100×600	$InCl_3$+Zn+ethanol, 150, 20h	/	[18]
	hollow spheres	100-150	$InCl_3$+Zn+ethanol, 120, 20h	/	
Fe Ni	nanoparticles	Fe-65 Ni-12	$Fe(acac)_3$ or $Ni(acac)_2$+N_2H_5OH+ethylenediamine, 140, 8h	Fe-bulk magnetic, Ni-superparamagnetion	[19]
Ni	nanobelts	15×150 ×2000	$NiCl_2.6H_2O$+$Na_2C_4H_4O_6$+ NaOH+SDBS+$NaH_2PO_2.H_2O$, 110, 8h	Ferromagnetic behavior	[20]
Ru	nanoparticles	2-10	$RuCl_3.xH_2O$+ethanol or methanol, 175, 24h	/	[21]
Se	nanorods	50×10000	Na_2SeO_3+PVA+DH_2O, 180, 20h	/	[22]
Te	nanowires	10 ×5000	Na_2TeO_3+glucose+SDBS+H_2O, 180, 12h	/	[23-25]
	Porous nanotubes	100×4000	Te+KOH+MCM-41+DMF, 180, 24h	/	
	Nanobelts, nanotubes	8×30×300 5×150×5000.	Na_2TeO_3+$NH_3.H_2O$, 180, 36h	/	

Table 1-2. The Experimental Conditions and Results of C Nanostructures

samples	*Type of nanostructure*	*Size (nm)*	*Starting materials & temperature (℃) & time*	*Properties (Raman cm^{-1})*	*References*
1	nanotubes	15×50×1000	Ethanol+NiO; 550	/	[26]
2	MW nanotubes	20×40×1200	Hexachorobenzene+K+benzene +Co/Ni; 350	1350, 1580	[27]
3	nanosheets	10×200×200	Tetracholomethane+K; 60-100	1321, 1576, 2639	[28]
4	nanotubes	4×20×60000	Hexachlorobenzene+Na+ $NiCl_2$+cyclohexane; 230	1370, 1604	[29]
5	nanofibers	70×90×700	Ethylene glycol+H_2O+Ni; 770	/	[30]
6	MW nanotubes	6×50×500	Amorphous carbon+H_2O; 800	1581	[31]
7	nanocells	Ø<100	Amorphous carbon+H_2O; 600	1350, 1600	[32]
	Nanocells	Ø<100	Amorphous carbon+H_2O; 600	1600	[33]
	MWnanotubes	4×10×500	Amorphous carbon+H_2O; 800	1581	
8	Amorphous C		C_{60} +H_2O;400	/	[34]
	nanotubes	20×40×120	C_{60} +H_2O;700	1608	
9	MWnanotubes	18×20×70	Polyethyele+H_2O; 800	1580, 1620, 2700	[35]

2.2. Effect of Experimental Conditions on Results

The experimental conditions, such as starting materials, solvents, surfactants, reaction temperature, and reaction time, have a crucial effect on the size, shapes, and properties of the nanocrystals. These effects will be discussed in the following sections.

2.2.1. Effect of Starting Materials and Solvent on Results

Different starting materials may have a different effect on the morphology and size of nanocystals. The first case is that the starting materials may have no obvious effect on the morphology and size of as-prepared nanostructures. One example is Ag nanostructures (Table 1-1). When the starting materials are silver octanethiolate and silver dodecanethiolate, the size of Ag nanodots is almost the same, which indicates the starting materials have no obvious influence on the shape and size of the obtained product. The second case is that the starting materials may have an important effect on the shape and size of obtained product. Comparing the following reaction system of silver octanithiolate (or silver dodecanethiolate) + water (system I), and the reaction system of $AgNO_3$+PVP+EG (system II), one can see Ag nanodots can be obtained in system I and Ag dendritic nanostructures can be prepared in system II under the same reaction temperature (180°C) and reaction time (12h) [11, 13] (Table 1-1), which indicates starting materials can affect the shapes of the resulted Ag nanostructures. The third case is that the surfactants and the concentrations of reactants have crucial roles on the morphology and size of prepared nanostructures. The example is Cu nanowires (Table 1-1). Cu nanowires formed with 3mM SDBS and irregular aggregated particles appeared without SDBS. When other surfactants, such as SDS (sodium dodecyl sulfate), PVP, CTAB (cetyltrimethylammonium bromide), were used instead of SDBS, no nanowires were produced. Low concentration of NaOH (<4M) and glycerol (25%) aggregated Cu particles. High concentration of both NaOH solution (~5M) and glycerol (~40%) led to the formation of Cu nanowires with high yield. However, the very high concentration of NaOH (7M) and glycerol (55vol%) also resulted in the aggregated particles. The fourth case is that the solvent may affect the morphology of the prepared nanostructures. The effect of the solvent on the shape can be illuminated by Sb nanostructures (Table 1-1). Keeping other experimental conditions identical, the as-prepared Sb nanostructure is tube-groove-like using methanol as the solvent and the obtained Sb nanostructure is dentritite using EG (ethyl glycol) as the solvent.

2.2.2. Effect of Reaction Temperature and Time on Results

Usually higher temperature and longer time lead to larger size of particles and good crystallization. Lower temperature and shorter reaction time result in smaller size and poorer crystallization. The reaction temperature and time also can affect the morphology of products. For the reaction temperature effect, the obvious examples are that (Table 1-1) In nanotubes could form at 150°C and In nano-hollow spheres formed at 120°C while the other conditions were kept identical, and that amorphous C (Table 1-1) obtained at 400°C and C nanotubes were prepared at 700°C when the starting materials and reaction time are identical. Compared sample 6 with sample 7 in table 1-2, we could see MW C nanotubes formed at 800°C and C nanocells and nanotubes obtained at 600°C when other experimental condition remained unchanged. The reaction time effect can be illustrated by Ag nanostructures [Table 1-1].

When the reaction time is 3h keeping other conditions identical, the obtained product is Ag chainlike nanocrystals. However, when reaction time is 4h, 8h, and 12h, the obtained product is Ag dendritic nanocrystals.

The effect of solvents, pH, and surfactants on the morphology of nanostructures can also be seen from Table 1-1. Most of the elemental nanostructures prepared in a pH>7, surfactants, and chain-like alcohol solvents are one-dimensional structures. For Ag nanodots, the solvent is water and no pH was adjusted in the experimental process. For Ag chainlike and dentritic nanostructures, Cu nanowries, Sb nanowires and tube-grooves, Bi nanotubes, In nanotubes, the solvents are alcohol. It seems solvent alcohol is beneficial to the growth of 1D nanostructural elements. For Sb nanotubes, Bi nanowires, Te nanobelts and nanotubes, NaOH, EN (ethylenediamine), and $NH_3.H_2O$ were used as pH adjustment. Actually, we could see from 1D nanostructures of other compounds, the proper basic solution favors the formation of 1D nanostructures. For the preparation of Ni nanobelts, Se nanorods, and Te nanowires, SDBS, PVA, and SDBS were used as surfactant to obtain the 1D nanostructures. We will see from other section that the certain surfactant has similar effect on the shape of some different nanomaterials. The exception is Fe, Ni, and Ru nanoparticles, the solvents are alcohol, but no 1D nanostructures were obtained. Another exception is carbon nanostructures. The 1D carbon structures could prepared without pH adjustment, alcohol solvents, and surfactants. For carbon nanostructures, there are plenty of reports. In table 1-2, we select these with well-dispersion, special shapes and size less than 20nm except carbon nanofibers.

2.2.3. Effect of Size on the Properties of Nanostructures

The size of nanocrystals has an important effect on their properties. One could see from Table 1-1, 4nm of Ag nanodots have UV-Vis absorption at 423nm. However, the larger size of Ag nanostructures has no absorption. Cu nanowires have a good absorption at 590nm. However, other metal nanostructures, such Sb, Bi, In, Ni, and Ru, and non-metal elements such as Se and Te, do not have absorption and photoluminescence. 65nm of Fe has bulk-like iron magnetic property and 12nm of Ni has superparamagnetion. The increasing of magnetocrystalline anisotropy is attributed to the surface effect of nanoscale nicked particles. For carbon nanostructures, the Raman was studied. Comparing sample 2, ample 6, sample 7, and sample 9, their shapes are MW nanotubes (multi-wall nanotubes). Their Raman peaks are almost at same position (1580 and 1581cm^{-1}). This indicates the nanostructures may have the same properties when they have same morphology and similar size.

3. Hydrothermal/Solvothermal Route to Oxide and Hydroxide Nanostructures

3.1. Introduction

Oxide and hydroxide nanostructures obtained from hydrothermal and solvothermal technique, cover a variety of groups. They include main metal oxides and hydroxides such as $Mg(OH)_2$ and MgO, $In(OH)_3$, SnO, SnO_2 and SiO_2, and Sb_2O_3, rare metal hydroxides, and transition metal oxides and hydroxides such as CuO, ZnO and CdO, VO_2 and V_2O_5, Ta_2O_5, Fe_2O_3 and Fe_3O_4, Co_3O_4, $Ni(OH)_2$, MnO_2, Mn_2O_3, and TiO_2. These compounds have different properties and applications. IIA metal hydroxide $Mg(OH)_2$ is used for medical and industrial products in broad fields. MgO nanostructures are expected to have many novel mechanical, catalytic, and electronic properties. They can be found many applications in functional composites as reinforcing agents and artificial pinning centers [36]. IIIA metal hydroxide $In(OH)_3$ and IVA metal oxide SnO_2 are semiconductors. SnO_2 is well known for its application in gas sensors [37-38], dye-base solar cells [39], electrode materials [40] and catalysts [41]. VA metal oxide Sb_2O_3 is a very attractive material because of applications as catalysts and gas sensors [42-43]. Sb_2O_3 is very useful as a high-efficiency flame-retardant synergist in plastics, paints, adhesives, and textile back coatings [44]. IIB metal oxide CdO and ZnO are semiconductors. CdO has the applications in solar cells [45-46], photodiodes [47], and gas sensors [48]. ZnO is a good candidate for blue-UV emission optoelectronic devices [71]. Rare earth metal hydroxide $La(OH)_3$ has long been used as the catalyzer and sorbent [83] and other rare earth compounds have been extensively used in high-performance luminescence devices, magnets, catalysts, and other functional materials because of their electronic, optical, and chemical characteristics resulting from the 4f shell of their ions [49]. IVB metal TiO_2 is of great interest for possible application in photovoltaic cells [50], semiconductor photo-catalyst [51], catalyst support [52], and gas and humidity sensor [53]. VB metal oxide VO_x and Ta_2O_5 can be used as catalysts [54]. VIIB metal oxide MnOx and VIIIB metal oxide FeOx, CoOx, NiO are known to be an active catalyst and as media for magnetic data storage, for solar energy conversion, and in the preparation of ferrofluids [55-59].

3.2. Effect of Experimental Conditions on Nanostructures

The nanostructures of oxides and hydroxides include 0-dimensional (0D) nanodots, 1D nanorods, nanotubes, nanowires, and nanofibers, 2D nanosheets and nanobelts, and 3D nanocubes and nano-spherical particles. Different experimental conditions have a various effect on the shapes and size of oxide and hydroxide nanocrystals. These experimental parameters include starting reactive materials, solvents, surfactants, acidity and basicity, reaction temperature, and reaction time. The experimental parameters and experimental results of oxides and hydroxides are listed in Table 2-1 and Table 2-2.

3.2.1. Effect of Starting Materials on Nanostructures

Starting materials include the starting reactive materials, solvents, acid and base, surfactants. All these experimental parameters influence the size, shape, and the other structures. Gao and his coworkers [63] reported 42×197nm of SnO_2 nanorods could be prepared in the solution of $SnCl_4$ and NaOH at 180°C for 24h but the obtained SnO_2 was 5×19nm when CTAB (cetyltrimethylammonium bromide) was added to the reaction system keeping other experimental parameters unchanged (Table 2-1). Keeping other conditions identical, Samulski et al [73] prepared 25×100nm and 40×500nm of ZnO nanorods from $ZnAc_2$ in methanol and ethanol, respectively (Table 2-1). These two results indicate the effect of starting materials surfactant and solvent on the size of nanostructure. In some case, the resultant could take the size and shape of the reactant. One could see from Table 2-1 that the size and shape of MgO nanorods is the same as those of reactant $Mg(OH)_2$. Actually, the starting materials can also affect the structure of the obtained nanomaterials. For example, α-

MnO_2 nanowires with size of 5×5000nm could be prepared from $MnSO_4$ and $(NH_4)_2S_2O_8$ at 120°C for 12h. β-MnO_2 could be obtained at the same experimental conditions but adding $(NH_4)_2SO_4$ in the above reaction system (Table 2-2). Besides solvents, surfactants, and the type of the starting reactive materials, the concentration of the starting reactive materials and pH affect the structures of nanocrystals as well. Hu et al [64] reported the size of SnO_2 nanorods decreased with the decreasing of the concentration of $SnCl_4$. On the other hand, the size of SnO_2 nanorods is 120×2000nm when the concentration of $SnCl_4$ is 0.06M. The size of SnO_2 nanorods decreased to 70×2000nm and 40×2000nm when the concentration of $SnCl_4$ became 0.003M and 0.0015M. When the concentration of $SnCl_4$ decreased to less than 0.0015M, the size of SnO_2 nanorods do not dramatically vary. However, the size of the nanocrystals does not always decrease with the decreasing of the concentration of reactants. For example, Fe_3O_4 nanowires could be prepared from Fe_3O_4 nanoparticles and NaOH [84] (Table 2-1). When the concentration of NaOH is 1.5M, 5M, and 10M, the size of Fe_3O_4 nanowires is 50×7000nm, 20×7000nm, and 15×7000nm, respectively. In this case, the size decreases with the increasing of the concentration of reactant because the NaOH is reactant and pH regulator. The concentration of reactant may affect the resultants. Qian et al studied the reaction between $KMnO_4$ and ethanol in aqueous solution. Keeping other conditions identical, when the volume percent of ethanol in the water (%) is 50-100, 30-50, 10-30, 2-10, 1-2, the obtained product at 160°C is Mn_3O_4, Mn_3O_4, Mn_3O_4+MnOOH, MnOOH, and K_2MnO_4, respectively. As for the effect of pH on the structure of nanocrystals, the preparation of ZnO nanostructures is a good case. Yang et al [76] reported the shapes of ZnO prepared at 120°C were particles (pH=12), flower-like consisting of nanorods (pH=13.5), flower-like (pH=7), and flower like consisting of nanorods (pH=9) when keeping other conditions identical. Acid also has different effect on the morphology and size of the prepared product [101-102]. From Table 2-1 and Table 2-2, one could see the shape of obtained different materials is 3D structure when water or other neutral organic solvents were used in absence of surfactants. However, most of the obtained oxides and hydroxides have one-dimensional (1D) structures when reactive systems contain basic solution.

3.2.2. Effect of Reaction Temperature and Time on Nanostructures

The effect of the reaction temperature and time on the structures of nanocrystals may act in a variety of aspects. First, reaction temperature and reaction time have no obvious effect on the shape and size. An example is that the shape of $Mg(OH)_2$ and MgO did not change much [60-61] with the increasing of reaction temperature and time. The crystalline intensity of $Mg(OH)_2$ and MgO increased with the increasing of temperature. Second, the reaction temperature has an obvious effect on the shape of prepared product. For the preparation of $In(OH)_3$ nanocrystals [62] in the presence of CA (citric acid), when the reaction temperature is 160°C and 200°C, the shape of obtained $In(OH)_3$ is nanocubes and spheres, respectively. Reaction temperature also has a distinct effect on the morphology of ZnO nanocrystals [74]. Using 5nm Zn nanoparticels and $ZnAc_2$ as starting materials and keeping other reaction conditions identical, when the hydrothermal reaction temperature is 220°C the as-prepared ZnO is nanocoral-reef; when the reaction temperature is 230°C the morphology of the obtained is nanofibers (Table 2-1). Third, the high reaction temperature and long time increase the size of as-prepared nanocrystals. In case of SnO_2, the size of SnO_2 nanocrysatls increased from 4.5nm to 5.4nm to 6.0nm when the temperature increased from 100 to 150 to

200°C while keeping reaction time for 24h and other conditions identical. Keeping the temperature at 150°C, when the reaction time increased from 24h to 72h, the size of the SnO_2 nanoparticles increased from 5.4nm to 5.8nm [66]. As for Fe_3O_4 nanoparticles [84], the reaction temperature and time have the similar effect on the size of the product. Keeping concentration of NaOH at 10M, when the temperature is 110°C, the size of Fe_3O_4 is 15nm, when the temperature is 160°C and the size is 30nm. The case is similar toTiO_2 [98-99]. In some cases, the resultant can not formed in lower temperature and shorter time. For example, the temperature and time have to be at least 180°C and 7 days for the formation of V_2O_5 nanotubes from $V_2O_5.nH_2O$ gel and hexadecylamine. There are other impurities with V_2O_5 nanotubes when temperature is lower than 180°C and reaction time shorter than 7 days.

3.2.3. Effect of Pre-Treatment on Nanostructures

Before hydrothermal/solvothermal reaction, the pre-treatment of starting materials may have a different influence on the product. This can be illuminated by the preparation of Co_3O_4 hollow spheres [87]. The suspension of the same starting materials was divided into three parts which were treated in three ways. Suspension A was stirred with an electromagnetic stirrer at room temperature for 0.5h. Suspension B was refluxed at 64-65°C (the boiling point of the solvent absolute methanol) for 0.5h. Suspension C was subjected to an ultrasonic agitation. Then the suspension A, B, and C were put into three Teflon-liner autoclaves, respectively and were treated under the same conditions. The morphology of the as prepared Co_3O_4 is hollow spheres but the size is quite different. The size of sample A hollow spheres is 200nm. Sample B is 70nm and the sample C is 50nm.

The pre-treatment also has an obvious influence on the morphology of nanocrystals, which will discuss in the other section.

3.2.4. Effect of Morphology and Size on the Properties of Nanostructures

Oxides and hydroxides have different properties. Here we limit to their optical properties. The absorption and photoluminescence (PL) peaks may have a blue-shift with the decreasing of the nanoparticles. One can compare the absorption and PL spectra of the CdO nanocrystals with 18nm, 9.5nm and 4.5nm [70]. The absorption of CdO nanocrystals is 2.3eV for 18nm, 2.8eV for 9.5nm (they are corresponding to bulk sample 2.5eV), and 3.27eV for 4.5nm. PL of the CdO nanocrystals is 540nm (560nm, 590nm) for 18nm nanocrystals and 495nm (540nm, 570nm) for 4.5nm crystals under the excitation 430nm (480nm, 520nm). The large blue-shift to the bulk CdO with 4.5nm nanocrystals is attributed to quantum confinement. The effect of shape on the PL is complex. Taking Sb_2O_3 nanopowder (starting materials) and nanotubes as examples [69] (Table 2-1), their PL peak and position are basically identical under the same excitation wavelength. But the intensity of Sb_2O_3 nanotubes is much stronger than that of Sb_2O_3 nanopowder. For TiO_2 nanowires (40×15000nm) and TiO_2 nanoparticles (starting materials), the PL peak shape and position are identical under the same excitation wavelength. But the intensity of TiO_2 nanowires is stronger than that of TiO_2 powder (Table 2-2).

Table 2-1. The Experimental Conditions and Results of Oxides and Hydroxides Nanostructures

Samples	Type of nanostructure	Size (nm)	Starting materials & temperature (°C) & time	properties	reference
$Mg(OH)_2$	Nanotubes	50×80×70000	$Mg_{10}(OH)_{18}Cl_2.5H_2O$+EN, 120, 6h	/	[60-61]
$Mg(OH)_2$	Nanorods	30×30000	$MgSO_4+NH_3.H_2O$ pH=12, 160, 10h		
MgO	nanorods	30×30000	$Mg(OH)_2$ nanorods, 360, 2h		
$In(OH)_3$	nanocubes	70×70×70	$InCl_3$+NaOH, 160, 20h	/	[62]
SnO_2	nanorods	42×197, 5×19, 4×39	$SnCl_4+NaOH+H_2O$+ethanol,180,24h $SnCl_4$+NaOH+CTAB+H_2O+ethanol,180,24h $SnCl_4$+NaOH+ethanol, 200, 8h	Absorption 345nm	[63-67]
	Nanorods	40×1200	$SnCl_4$+NaOH+CTAB+H_2O,160,12h	/	
	nanodots	3	$SnCl_4$+urea+CTAB+n-pentanol+ n-hexane, 150, 15h	Gas sensing	
	nanoparticles	3-4	SnO_2 precipitate, 100, 24h	Gas sensor	
	nanoparticles	4-5	$SnO_2.nH_2O$, 250, 5h	Gas sensor	
Quartz	nanocrystals	10>	Amorphous silica+NaOH,300, 2h	/	[68]
Sb_2O_3	nanotubes	30×40×7000	$SbCl_3$+CTAB+$NaBH_4$+ethanol,150,30h	PL 540nm	[69]
CdO	nanocrystals	4.5	$Cd(cup)_2$+TOPO+toluene, 240, 2.5h	PL495,540,570	[70]
CuO		10	$Cd(cup)_2$+TOPO+toluene, 180, 5h	/	
ZnO	nanowires	50×6000	$Zn(NO_3)_2+(NH_4)_2CO_3$+PEG, 200, 10h	PL 393	[71-76]
	nanowhiskers	150×20000	Zn+ ethylenediamine, 220, 10h	PL382	
	Nanorods	25×100, 40×500	$ZnAc_2$+methanol, 150, 24h $ZnAc_2$+ethanol, 150, 24h	/	
	Nanocoral-reef	/	5nm Zn nanoparticles+$ZnAc_2$220,24h	PL 394	
	nanofibers			PL 392	
		500×500000	5nm Zn nanoparticles+$ZnAc_2$330,24h		
	Nanobelts	10×20×2500	$ZnSO_4$+carbamide+ethanol+NaOH,160,12h	PL 3.23-3.3eV	
	Flower-like	/	$ZnAc_2$+NaOH or $NH_3.H_2O$, 120, 20h	/	

Table 2-1. The Experimental Conditions and Results of Oxides and Hydroxides Nanostructures (continued)

Samples	*Type of nanostructure*	*Size (nm)*	*Starting materials & temperature (°C) & time*	*properties*	*reference*
V_2O_5	*Nanotubes*	$30\times50\times500$	$V_2O_5.nH_2O$ *gel+hexadecylamine, 180, 168h*	/	*[77-78]*
VO_2	*nanobelts*	$10\times50\times5000$	NH_4VO_3*+formaic acid, 180, 48h*	*magnetization*	
Ta_2O_5	*nanoparticles*	*15*	*Tantalum pentabutoxide, 300, 2h*	*Photocatalytic*	*[79]*
$La(OH)_3$- $Ni(OH)_2$	*Nanorods nanocables*	25×700 coating 10-20	$La(NO_3)_3+Ni(NO_3)_2+NaOH$*, 180, 24h*	/	*[80]*
$La(OH)_3$	*nanorods*	10×300	$La(NO_3)_3+KOH$*, 200, 16h*	/	*[81]*
$La(OH)_3$	*Nanorods*	10×600	$La(NO_3)_3 + NH_3.H_2O$*, 210, 3h*	/	*[82]*
$Tb(OH)_3$& $Y(OH)_3$	*nanotubes*	$15\times30\times1000$	Tb_4O_7 *powder +NaOH, 170, 48h* Y_2O_3 *powder +NaOH, 170, 48h*	/	*[83]*

Table 2-2. The Experimental Conditions and Results of FeOx, Co_3O_4, $Ni(OH)_2$, MnOx, and TiO_2 Nanostructures

samples	Type of nanostructure	Size (nm)	Starting materials & temperature (°C) & time	properties	reference
Fe_3O_4	Nanowires	20×7000	Fe_3O_4 nanoparticle+NaOH,110,24h	Magnetism	[84-85]
	nanoparticles	8-11	Fe(acac)3+PVP+EG+hydrazine,180,12h	superparamagtic	
γ-Fe_2O_3	nanoparticles	10	Fe(cupf)3+n-octylamine+benzene,220,1h	magnetism	[86]
Co3O4	Hollow sphere	50	$Co(NO_3)_2$+methanol+SDBD, 180, 4h	/	[87]
$Ni(OH)_2$	nanorods	25×500	$Ni(NO_3)_2$+NaOH+AAO, 150, 24h	/	[88]
α-Mn_2O_3	nanorods	60×700	MnO_2+ammonia, 250, 24h	/	[89]
α-MnO_2	Nanowires	5×5000	$MnSO_4$+$(NH_4)_2S_2O_8$, 120, 12h	/	[90]
β-MnO_2	nanowires	40×2500	$MnSO_4$+$(NH_4)_2S_2O_8$+$(NH_4)_2SO_4$,120,12h		
γ-MnO_2	Urchinlike	40×1000	$MnSO_4$+$KBrO_3$, 130, 16h	/	[91]
	sisallike	20×10000	$MnSO_4$+$KBrO_3$+PEG, 110, 16h		
Mn_3O_4	nanocubes	8-16	$KMnO_4$+ethanol, 160, 8h	magnetism	[92]
MnO_2	nanobelts	5×1000	Mn_2O_3 powder + NaOH, 170, 12h	/	[93]
TiO_2	nanotubes	~10×75×10000	TiO_2 powder+10M NaOH, 120-190, 22h	/	[94]
		~10×15×1000	TiO_2 powder+10M NaOH, 110-150, 24h		[95]
		~5×10×1000	TiO_2 powder+10M NaOH, 120-150, 72h		[96]
	nanowires	40×15000	TiO_2 powder+10M NaOH, 200, 24h	PL545nm	[97]
	nanorods	9×100	$TiCl_4$, 50-150, 12h	/	[98]
	nanoparticles	5-25	$TiCl_4$+HCl, 80-150, 3-70h	Absorption303nm	[99]
	nanoparticles	20	Titanium isopropoxide + benzene, 250, 3h	/	[100]
	nanospindles	80×300	Tetrabutyl titanate + surfactants, 120, 13h	photocatalytic	[101]
	nanoparticles	20	Titanium n-butoxide + acids, 100-220, 10-72h	/	[102]
	Porous film	25	$TiCl_4$, 100, 12h	Solar cells	[103]
	Porous nanoparticles	~60-100(7-27)	$H_2TiO(C_2O_4)_2$, 150-250, 1-6h	/	[104]
TiO_2-Er	Nano-spindles	20×70	$Ti(OC_3H_7)_4$+$N(CH_3)_4OH$, 210, 3h	PL 1532nm Colloidal plate	[105]
TiO_2	nanobelt	100×800	$TiCl_4$+10M NaOH, 250, 6h	Li- storage	[106]

Table 3-1. The Experimental Conditions and Results of Chalcogenide Nanostructures

Samples	*Type of nanostructure*	*Size (nm)*	*Starting materials & temperature (℃) & time*	*Properties*	*References*
SnS	*Nanorods*	*10×55*	$SnCl_2$+*thiourea, 180, 2h*	/	*[129]*
	nanoparticles	*3-10*	$SnCl_2$+*thiourea, 180, 8h*		
SnS_2	*Nano-belts*	*80×80000*	$SnCl_2$+Na_2S+*TGA, 200, 20h*	/	*[130-131]*
	nanoflakes	*2×250×500*	*Tetrabutyltin*+CS_2*, 200, 40h*	*Raman312cm*$^{-1}$*, Mössbauerδ1.002*	
PbS	*nanorods*	*30×50000*	$PbAc_2$+*En*+CS_2*(benzene), 160, 12h*	/	*[132]*
	frame,	*Ø=10-20um*	$Pb(NO_3)_2$+*thhiourea+polyamide,140,20h*	/	*[133]*
	nanopearls	*100×5000*	$PbCl_2$+CH_3CSNH_2*, 120, 18h*	/	*[134]*
	Nanoparticles	*5-7*	$Pb(NO_3)_2$ +Na_2S*, 20, 12h*	/	*[135]*
	Nano-cubes	*20×5000*	$Pb(NO_3)_2$ +Na_2S*, 120, 12h*		
	nanowires	*300*	$Pb(NO_3)_2$ +*thiourea, 120, 12h*		
	dendritic	*40×20000*	$PbAc_2$ Na_2S+*CTAB, 120, 24h*	/	*[136]*
	cube	*22*	$PbAc_2$+$Na_2S_2O_3$+$C_{37}H_{33}COOK$*,180, 24h*	/	*[137]*
CaS & SrS	*Nanoparticles*	*20*	$CaCl_2$ *or* $SrCl_2$+*S+EN, 170, 12h*	*PL 550nm*	*[138]*
	Plate-like	*20*	$CaCl_2$ *or* $SrCl_2$+*S+EN, 220, 12h*	*PL 465nm*	
Sb_2S_3	*nanorods*	*30×500*	$SbCl_3$+Na_2S+H_2O*, 180, 12h*	/	*[139-140]*
		60×450	$SbCl_3$+*thiorean, 120-180, 2-12h*		
	Nanorods	*40×10000*	$SbCl_3$+CS_2*, 160, 5h*	/	*[141]*
	Dendrite-like	/			
	Straw-tie-like	/			
	nanowires	*20×50000*	$SbCl_3$+Na_2S*, 200, 10h*	/	*[142]*
Sb_2Se_3	*Nanowires*	*30×8000*	$SbCl_3$+Na_2S+*Se, 120, 60h*	/	*[143-144]*
	nanoribbons	*25×30000*	$SbCl_3$+Na_2SeSO_3+*PVA*+$NH_3.H_2O$ +*tartaric acid, 180, 20h*	*UV-Vis 727nm*	

Table 3-1. The Experimental Conditions and Results of Chalcogenide Nanostructures (continued)

Samples	Type of nanostructure	Size (nm)	Starting materials & temperature (℃) & time	Properties	References
Bi_2S_3	nanorods	10×1000	$Bi(NO_3)_3$+Na_2S, 110, 12h	/	[145-146]
		45×500	$Bi(NO_3)_3$+Na_2S+EDTA, 95, 12h		
	Flower-like	30×400 (rods)	$Bi(NO_3)_3$+$Na_2S_2O_3$, 200, 16h	/	[147]
	nanowires	30×5000	$Bi(NO_3)_3$+Na_2S+TGA, 200, 16h	/	[148]
	Nanoribbons	20×100×500000	$Bi(NO_3)_3$+ $Na_2S_2O_3$+NaOH+glycerol,24h	/	[149]
		10×65×20000	$Bi(NO_3)_3$+ Na_2S+NaOH+glycerol, 72h		
FeS_2	Nanorods	60×700	$Fe_2(SO_4)_3$+thiourea, 180, 12h	/	[150]
	Nanotubes	60×120×700	$Fe(NO_3)_3$+thiourea, 180, 12h		
	nanoribbons	100×10000	$Fe(NO_3)_3$+thiourea, 180, 12h		
$NiSe_2$	nanorods	30×120	$NiCl_2$+Se, 160, 10h	/	[151]
$NiTe_2$		20×90	$NiCl_2$+ Te, 180, 15h		
α-MnS	nanorods	40×500	Mn +S powder, 240-260, 14-20h	PL 400	[152]
MoS_2	nanowires	4×50	MoO_3+Na_2S+HCl, 260, 12h	/	[153]

Table 3-2. The Experimental Conditions and Results of Cu (II) Zn (II) Cd (II) Chalcogenide Nanostructures

Samples	Type of nanostructure	Size (nm)	Starting materials & temperature (°C) & time	Properties	References
CuS	nanoparticles	12	CuCl+ thiourea, 110, 40h	Absorption 515nm 555nm	[154]
	Nanowires	40×500	CuCl+ thiourea+TEDA, 110, 40h		
	Nanotubes	20×30×400	CuCl+thiourea+TMEA,110, 40h		
	nanovesicles	50	CuCl+thiourea+DBA,110, 40h		
	Nanospheres	550	0.001mol $CuSO_4$+$Na_2S_2O_3$ 150, 12h	/	[155]
	nanotubes	40×200×20000	0.005mol $CuSO_4$+$Na_2S_2O_3$ 150, 12h		
	urchin-like	70	0.0025mol $CuSO_4$+$Na_2S_2O_3$150,12h		

Table 3-2. The Experimental Conditions and Results of Cu (II) Zn (II) Cd (II) Chalcogenide Nanostructures (Continued)

Samples	Type of nanostructure	Size (nm)	Starting materials & temperature (°C) & time	Properties	References
$ZnS-Mn^{2+}$	nanospheres	100	$ZnCl_2$+$MnCl_2$+ olec acid, 180,60h	PL 572, 450nm	[156]
$ZnS-Al^{3+}-Cu^{2+}$			$ZnCl_2$ + $CuCl_2$+$AlCl_3$+olec acid, 180, 60h	PL 530nm	
ZnSe	nanorods	20×1000	$Zn(NO_3)_2$+Se+nbutylamine,220,12h	/	[157]
CdS	Nanoparticles- polymer	13	$CdCl_2$+thioacetamide+VAC+AIBN,80,12h	/	[158]
	rods	70×700			
	nanocrystals	5	Cadmium stearate+S, 220, 12h	Absorption 455	[159]
	Rods	40×8000	CdS nanoparticles+EN, 140, 5h	/	[160]
	Twinrods	6×50	CdS nanoparticles+S+EN, 180, 5h		
	tetrarods	40×1700	CdS nanoparticles+EN, 180, 5h		
	nanowhiskers	60×12000	$CdCl_2$+thiosemicarbazide, 170, 14h	/	[161]
	nanowires	30×3000	$CdCl_2$+thioureau+HQ+EN, 170, 50h	Absorption 505	[162]
	Nanorods	30×200-A	$CdCl_2$+thiourea+EN,160, 48h	PL 365	[163]
	Nanorods,twinrods,tripods	20×320-B	$CdCl_2$+thiourea+EN+1.81C_{12},160, 48h		
	Nanorods	15×80-C	$CdCl_2$+thiourea+EN+1.67C_{14},160, 48h		
	nanorods	30×200-D	$CdCl_2$+thiourea+EN+1.67C_{16},160, 48h		
	Twi, tri, tetra nanorods	20×200	$CdSO_4$+Na_2S+TGA, 200, 16h	PL 445nm	[164]
CdS	Nanowires	40×100000	$CdCl_2$+thiourea+PAM, 170, 240h	PL	[165]
CdS-Se	Core-shell-nanowires	7.7×.75×1000	CdS nanowires+Se+TBP, 100, 24h	No PL	
CdE	CdS nanorods	10×200	Cd+S+EN, 120, 8h	Raman 260,	[166]
	CdSe nanorods	20×500	Cd+Se+EN, 200, 8h	$640cm^{-1}$	
	CdTe nanorods	10×500	Cd+Te+EN, 200, 40h	PL 467nm	
CdSe	nanoparticles	3	Cadmium stearate+Se+ tetralin+benzene	PL 540	[167-168]
		7	CdCl2+Se+Na+EN, 120, 6h	PL 600	
CdTe	nanocrystals	2-4	Cd^{2+}/TEG/NaHTe-=1:2.4:0.2, 100, 2h	PL green to red	[169]
			Cd^{2+}/TEG/NaHTe=1:2.4:0.2, 100, 2h		
			Cd^{2+}/TEG/NaHTe=1:2.4:0.2, 100,2h		

4. Hydrothermal/Solvothermal Preparation of Nanostructures of Chalcogenides

4.1. Introduction

Metal chalcogenides have been extensively studied because of their potential application in electronic, optical, and superconductor devices [107-108]. Main-group metal chalcogenides such as AB_2 and AB (A=Ca, Sr, Sn, Pb; B=S, Se, Te) group have potential applications in phosphor materials [145], solar cell materials [109], holographic recording materials and electrical switching [110], in sensors, lasers, infrared detectors. The A_2B_3 (A=Sb, Bi; B=S, Se, Te) group compounds are semiconductors and have applications in television cameras with photo-conducting targets, thermoelectric devices, optoelectronic devices, and in IR spectroscopy [111]. Secondary-group metals chalcogenides such as AB (A=Fe, Ni, Mn, Cu, Zn, Cd; B=S, Se, Te) are solar cell materials [112], non-aqueous lithium batteries [113], catalytic hydrodesulphurization of petroleum [114-115], and wear resistance [116], photoluminescence materials, electroluminescence materials, mechano-luminescence materials, thermal-luminescence materials, infrared windows, sensors, and lasers [117-124], field-effect transistors, light emitting diodes, photocatalysis ans biological sensors [125-128].

4.2. Effect of Experimental Conditions on Nanostructures

The chalcogenide nanomaterials discussed here include IIA-VI compounds CaS and SrS, IVA-VI compounds such as SnS_2, SnS, PbS, and VA-VI compounds Sb_2S_3, Sb_2Se_3, and Bi_2S_3, IB-VI compound CuS, IIB-VI compounds ZnS, ZnSe, CdS, CdSe, and CdTe, VIB-VI compound MoS_2, VIIB-VI compound MnS, and VIIIB-VI compounds FeS_2, $NiSe_2$, and $NiTe_2$. The nanostructures of these compounds comprise nanoparticles, nanoflakes, nanospheres, nanocubes, nanorods, twinrods, tetrarods, tripod, nanopearls, nanowires, dendritic structure consisting of nanorods, nanoribbons, flower-like structure consisting of nanorods, and nanowhiskers. The formation of these structures was influenced by the internality of materials and preparing conditions. Here we will discuss the effect of experimental conditions on the nanostructures of chalcogenides. The experimental conditions and results of chalcogenides are listed in Table 3-1 and Table 3-2.

4.2.1. Effect of Starting Materials on Nanostructures

As we discussed in the previous sections, the starting materials including the reactant materials, surfactants, and solvents, have an obvious influence on the structure of resulting product. The effect of starting materials on the nanostructures could be illuminated by the formation of different nanostructures of PbS, FeS_2, CdS, Bi_2S_3, and Sb_2Se_3 (Table 3-1 and Table 3-2). PbS nanopearls consisting of PbS nanoparticles formed when $PbCl_2$ reacted with CH_3CSNH_2 under hydrothermal condition at 120°C for 18h. When Na_2S was used as sulfur sources instead of CH_3CSNH_2 keeping other conditions constant, the obtained PbS nanoparticles are similar but no nanopearls formed. When $PbAc_2$ was used instead of $PbCl_2$, the as-formed shape of formed PbS is irregular and cube-like [141]. Another example is the formation of PbS nano-cubes and nanowires [135]. Keeping $Pb(NO_3)_2$ as lead source, when Na_2S as sulfur source, the morphology of formed PbS is nano-cubes. When thiourea was used in stead of Na_2S under other conditions identical, the shape of obtained PbS is nanowires. The shape of FeS_2 was affected by the reactant as well [150]. When $FeSO_4$, $FeCl_3$, and $Fe(NO_3)_3$ were used as iron resource keeping other conditions constant, the corresponding shape of the obtained FeS_2 is short nanorods, nanotubes, and nanoribbon. In some case, adding another sulfur source into the reaction system has an important influence on the shape of the nanostructure. In a reactive system consists of CdS nanoparticles and EN, the obtained product after 5h at 180°C is CdS tetrarods. When sulfur powder was added to the same reactive system while keeping other conditions identical, the shape of obtained CdS is twinrods [160]. The amount of the starting reactant could also affect the shape and size of product. One could see from Table 3-2 that CuS nanospheres, CuS nanotubes, and CuS urchin-like structure consisting of nanotubes formed when the amount of reactant $CuSO_4$ is 0.001mol, 0.005mol, and 0.0025mol, respectively, while keeping other conditions identical [155]. The amount and type of the surfactant also have different influences on the nanostructures. In the formation process of PbS frame [133], the monomer of surfactant is acrylamide. When the amount of acrylamide is 0.018mol, 0.02mol, 0.022mol, the shape of frame is ellipse, hexagon, and tube, respectively. The dendritic nanostructure of PbS could form in the system of $PbAc_2$, Na_2S, and CTAB but only PbS cubes formed in the absence of CTAB surfactant keeping other experimental parameters constant [136]. Flower-like Bi_2S_3 nanorods were prepared with TGA (thioglycolic acid) but only nano-tips were observed without TGA assistance in the reactive system of $Bi(NO_3)_2$ and $Na_2S_2O_3$ [147]. Surfactant PVA and its amount have a distinct influence on the shape and size of Sb_2Se_3 [144]. When the molar ratio of PVA and Sb^{3+} was 6/1, the obtained Sb_2Se_3 was large particles and rods. When the ratio was 23/1, Sb_2Se_3 nanoribbons formed. As for the effect of the type of surfactants on the structure, one could see clearly from the preparation of CuS nanostructure in Table 3-2 [154]. In the reactive system of CuCl and thiourea, when no surfactant, TEDA (triethylenediamine), TMEA (tramethylethylenediamine), and DBA (di-n-butylamine) was used as surfactant while keeping other conditions identical, the morphology of as-prepared CuS is nanoparticles, nanowires, nanotubes, and nanovesicles, respectively. In the preparation of CdS nanorods, and twinrods and tripods, the type of surfactant has an obvious effect on the shape of CdS as well [163]. When no surfactant, CdS nanorods formed from $CdCl_2$ and thisurea in EN. When C_{12} (S-dodeclisothiounium) was acted as surfactant while keeping other conditions identical, CdS nanorods (40%), CdS twinrods (20%), CdS golfclub (10%), and CdS tripod-shape were obtained. In some case, the type of the surfactant has no distinct influence on the shape and size of formed nanostructure. An example is the preparation of CdS nanorods in the $CdCl_2$ and thiourea reaction system [163]. The shape of CdS kept the same and the size of nanorods had no big change when C_{14} (H_3C-(CH_2)n-Br, n=13) and C_{16} (H_3C-(CH_2)n-Br, n=15) were used as surfactant, respectively. Generally, solvents affect the shape and size of as-prepared nanostructure. Formation of CdS nanocrystals with different shapes from elemental reaction in EN (ethylenediamine), DEN (diethylamine), pyridine, and 1, 6-diaminohexane, was taken as an example. CdS in EN was long nanorods with diameter of 10nm. CdS in DEN is spherical nanoparticle with the average size of 20nm. CdS in pyridine is spherical nanoparticles with average size of 40nm. CdS in 1, 6-diaminohesane is tripod nanostructure. In the previous sections, what was discussed is one-phase reaction. In fact, the reaction could be controlled into two-phase in one reaction system through using hydrophilic and hydrophobic solvents. One example is the preparation of PbS nanostructures [132]. Keeping $PbAc_2$ in ethylene phase in the lower part of autoclave and CS_2- benzene phase in above part, the obtained PbS is nanorods. When the reaction processed in one-phase of ethylene, the as prepared PbS is nanocubes keeping other conditions identical [132]. The solvent even affected the formation of products [151]. $NiSe_2$ (and $NiTe_2$) nanorods could be prepared from $NiCl_2$ and Se (and Te) in EN (ethylenediamine). When benzene was used as the solvent instead of EN keeping other conditions identical, neither $NiSe_2$ nor $NiTe_2$ could be obtained. When DEN (diethyleneamine) was used, no pure $NiSe_2$ or $NiTe_2$ could be obtained.

4.2.2. Effect of Reaction Temperature and Reaction Time on Nanostructures

Usually higher reaction temperature and longer reaction time may result in the increasing of size. The solvothermal reaction of $SbCl_3$, elemental Se, and Na_2SO_3 lead to the formation of Sb_2Se_3. The size of Sb_2Se_3 prepared at 120°C is 20-50nm of diameter and 2-8um of length and the size of Sb_2Se_3 prepared at 140°C is 50-80nm of diameter and 10um of length [143]. However, reaction temperature and reaction time could change the shape of nanostructures. In the formation of CaS and SrS nanostructures [138], the CaS and SrS prepared at 220°C display a plate-like morphology with an average diameter of 60-100nm while the CaS and SrS prepared at 170°C are spherical particles with an average diameter of 20nm. The morphology of FeS_2 nanostructure also changed with temperature [150] (Table 3-1). Keeping other conditions identical, at 150°C FeS_2 nanoparticles, nanorosd and nanowires were obtained; at 180°C uniform FeS_2 nanowires with very high aspect ratio were produced; at 210°C FeS_2 nanoribbons with width of 100-250nm and length with up to tens of micrometer were prepared. At 230°C hexagonal nano-sheets along with some long rectangular sheets were obtained. For CdS nanostructures from $Cd(NO_3)_3$ and thiourea, the shape of CdS prepared at 160°C is irregular nanoparticles. At 220°C, the shape of obtained CdS is short nanorods with an aspect ratio of about 4-5. Reaction time could affect the morphology of chalcogenide nanostructures as well. For example, when reaction time is 2h and 8h keeping other conditions identical, the formed SnS from the hydrothermal reaction of $SnCl_2$ and thiourea is nanorods and nanoparticles, respectively [129]. Reaction time could only affect the length of nanorods and other 1D nanostructures but has no obvious effect on the diameter or even reduce the diameter. A distinguishing example is CdS nanorods [161] and Sb_2Se_3 nanoribbons [144]. The reaction time has no obvious effect on the width of CdS nanorods but has a distinct effect on the length of CdS nanorods [161]. CdS obtained from an experiment time of 7h with $CdCl_2$ and thiosemicarbazide in the molar ratio of 1/2 is short nanorods with the width of 50-70nm. When the reaction time was extended to 14h keeping other conditions identical, the CdS nanorods grew to lengths of more than 12um, but remained nearly unchanged in width. For Sb_2Se_3 nanostructurs, it is found the width of obtained Sb_2Se_3 nanoribbons from $SbCl_3$ and Na_2SeSO_3 is about 180nm and the length is about 5um when the reaction time is 3h. When the reaction time extended to 20h, the width of nanoribbons is about 50nm and the length is about 8um. Reaction time could affect the shape and the size at the same time. The case is the formation of Sb_2S_3 nanostructures [142]. When the reaction of $SbCl_3$ and Na_2S lasted for 1h, amorphous Sb_2S_3 nanoparticles with irregular shape were obtained. After 3h, some adjacent nanoparticles were joined with each other forming neck-like connections. When reaction time is 6h, well-crystallized Sb_2S_3 nanorods with the average size of 60×2000nm were produced. After 10h, the large quantities of uniform Sb_2S_3

nanowires with average size of 50×5000nm formed. It indicated the shape of Sb_2S_3 tended to develop into 1D direction with the extending of reaction time. The work of Qian [139-140] gave the similar results. Relative short reaction time could lead to the impurity chalcogenide product. In the reactive system of $Bi(NO_3)_3$, Na_2S, NaOH, and glycerol [149], when reaction time is 3h, amorphous $NaBiS_2$ formed; when reaction time is 24h, $NaBiS_2$ still coexists with expected Bi_2S_3 phase. It took over 72h to form pure Bi_2S_3 nanoribbons .

4.2.3. Effect of Pre-Treatment on the Nanostructures

The pre-treatment of starting materials had a different influence on the nanostructure of oxide Co_3O_4 which was discussed previously. It is found that the pre-treatment also had an obvious effect on the shape and size of chalcogenide Bi_2S_3 nanostructure [146]. Under stirred-assisted pre-treatment, the as prepared Bi_2S_3 is uniform single-crystal nanorods with an average diameter of 43nm. Under the oscillated-assisted pre-treatment, the obtained Bi_2S_3 are non-uniform nanorods with average diameter of 28nm. While under no pre-treatment, the formed Bi_2S_3 is polycrystalline nanorods with average diameter of 25nm.

4.2.4. Effect of Shape and Size of Nanostructures on the Properties

The extensively studied properties on chalcogenide nanocrystals are Raman, Mössbauer, photoluminescence, and absorption. The Raman of nanocrystals could be weaker than that of corresponding bulk materials. The Raman spectrum of SnS_2 nanoflakes [131] exhibited an intense peak at about 311cm^{-1} (A_{1g}). The peak at 210cm^{-1} of the first-order Eg of bulk materials disappeared, which probably resulted from the nanosize effect.

The PL emission of nanocrystals usually has a blue-shift compared to the corresponding materials. The smaller nanoparticles cause more blue-shift or more intense emission. For example, the emission peak of 20nm SrS nanoparticles is at 465nm which has an obvious blue-shift compared to that of bulk materials. In addition, the intensity of 20nm SrS is much stronger than that of 100nm SrS [138] (Table 3-1). However, for MnS nanorods (40×500nm), the emission is at 400nm, which is similar to the bandgap of the bulk counterpart [152]. The different shapes or ratios of length and diameter may lead to the different PL intensity. One could see from Table 3-2 [163] that A (nanorods), B (nanorods, twinrods, tripods), C (nanorods), and D (nanorods) have different ratios of length and diameter but they have a PL peak at 365nm with different intensity. The intensity of A and D is similar for the morphologies and size of A and D are similar. The intensity of C is the strongest among them for the size of C is the smallest. The PL intensity of B and C is different because of their different shapes and ratios of length and diameter. The PL peak of CdS with two-, three- and four-armed nanorods (20×200nm) is at 445nm [164]. This result is quite different from these of literature [170] probably because their shapes are different. For CdSe nanoparticles [167-168] in table 3-2, the PL peak of 3nm CdSe is at 540nm and the PL peak of 7nm CdSe is at 600nm. They obey the rule that the PL peak of smaller size nanoparticles shifts to lower wavelength. The size of nanoparticles could increase with the reaction time and reaction temperature. Therefore, the PL peak could shift to a higher wavelength (red shift) with the increasing of reaction time and reaction temperature. The case is CdTe nanoparticles [169]. The PL peak of CdTe shift from green to red light with the increasing of time and temperature.

The absorption of nanocrystals also could show a blue-shift compared to the corresponding bulk materials. The absorption peak of Sb_2Se_3 nanoribbons (25×30000nm) at

727nm (1.7eV) shows an obvious blue-shift relative to that of the bulk Sb_2Se_3 (1.3eV) [144]. For the nanoparticles with different size and shape, the peak and the intensity of absorption could be different. The absorption of CuS nanotubes (20×30×400nm-inner diameter× outer diameter × length) is at 515nm and the absorption of CuS nanovesicles (50nm) is at 555nm, and the intensity is weaker than CuS nanotubes [154]. For CdS nanoparticles [159], when the size of CdS is 4nm, the absorption peak is at 455nm; when the size is 5nm, the absorption peak is at 470nm; when the size is 10nm, the absorption peak is at 490nm. The absorption of bulk CdS is at 515nm. The absorption peak of CdS nanorods with size of 30×30000nm is at 505nm [162], which is a little blue-shift compared to the bulk CdS.

5. Hydrothermal/Solvothermal Preparation of Nanostructures of Ternary and Multi-Elemental Compounds

5.1. Introduction

So far, the ternarey and multi-elemental compounds prepared by hydrothermal/solvothermal route are Ag_3CuS_2, $CuFeS_2$, $CuInS_2$, $CdIn_2S_4$, $CdWO_4$, $BaTiO_3$, $PbSnS_3$, SbSI, $Ca_{10}(PO_4)_6(OH)_2$, $CoAl_2O_4$, $Ce_{1-x}ZrxO_2$, $La_{0.5}Ba_{0.5}MnO_3$, and $La_{0.5}Sr_{0.5}MnO_3$. All these compounds have interesting properties and potential applications. IB-VIIB-VI compound Ag_3CuS_2 has an average Vicher hardness of ca. 25.6, and has been found as a major component in the membranes of several commercial copper ion-selective electrodes [170]. IB-VIIIB-VI compound $CuFeS_2$ is an antiferromagnetic semiconductor with chalcopyrite structure [171-172]. It has novel optical absorption edge (0.5-0.6eV) [173], and a very high Neil temperature (550℃) [174]. IIB-IIIA-VI compound $CuInS_2$ also has a chalcopyrite structure and is a leading material for high-efficiency and radiation-hard solar cell applications [175]. In addition, $CuInS_2$ is a candidate for the cathode material of photochemical devices owing to its high performance and high output stability [176]. $CdIn_2S_4$ has good photoelectric properties [177]. IIB-VIB-VI compound $CdWO_4$ crystals have a monoclinic wolframite structure and are considered to be highly functional materials because of their high average refractive index, low radiation damage, low aftergrow to luminescence and high x-ray absorption coefficient [178]. As the V-VI-VII compound, SbSI is a ferroelectric material having many interesting properties [195] such as high photoconductivity, ferro-electricity and piezoelectricity. $Ca_{10}(PO_4)_6(OH)_2$ has been studied extensively and applied in a variety of fields. It is the mineral constituent of human hard tissues (bones, teeth, etc.) and is of importance in the biomaterials field as a raw material for the preparation of artificial bone graft [179-181]. It is also a promising material as reinforcing filter for composites, insulating agents, and chromatomedium for simple and rapid fractionation of proteins and nucleic [182-184]. $CoAl_2O_4$ is a double oxide with a normal spinel-type structure. It is well known as Thenard's blue for its impressive optical property and widely used in ceramics, glass, paint industry, and color TV tubes as contrast-enhancing luminescent pigment [185]. $Ce_{1-x}ZrxO_2$ solid solution is widely used in three-way catalyst formation for clean up of automotive gases [186]. $La_{0.5}Ba_{0.5}MnO_3$ and $La_{0.5}Sr_{0.5}MnO_3$ have the potential field-sensor, magetoresistive effect and device applications [187, 210].

Table 4. Experimental Conditions and Results of Ternary and Multi-elemental Compound Nanostructures

Samples	Type of nanostructure	Size (nm)	Starting materials & temperature (℃) & time	Properties	References
Ag_3CuS_2	Nanorods	30×200	AgCl+CuCl+thiourea, 150, 8h	/	[188]
$CuFeS_2$	Nanorods	20×5000	$CuCl+FeCl_3+(NH_4)_2S$, 200, 10h	/	[189]
$CuInS_2$	nanorods	20×400	$CuCl_2+In+CS_2+NaOH$, 180, 15h	Absorption 825	[190]
$CdIn_2S_4$	nanorods	10×200	CdS nanorod+$InCl_3$+thiourea+EN, 180, 10h	Raman 130, 300	[191]
$CdWO_4$	nanorods	20×80	$CdCl_2+NaWO_4$, 130, 5h	PL 486	[192]
$BaTiO_3$	nanoparticles	77 ± 23	$BaCl_2+TiCl_4$+KOH+Tween80, 230, 0.5h	Raman 300, 520, 716	[193]
$PbSnS_3$	nanorods	50×8000	$PbCl_2+SnCl_4+NH_2CSNH_2+I_2$, 200, 10h	Raman 201, 309, 442	[194]
SbSI	nanorods	20×2000	$SbCl_3+(NH_2)_2S+I_2$, 180, 8h	/	[195]
$Ca_{10}(PO_4)_6(OH)_2$	nanorods	10×1500	$Ca(NO_3)_2+Na_3PO_4$+CTAB, 150, 10h	/	[196]
$CoAl_2O_4$	nanocubes	70	$CoCl_2+AlCl_3$+3M NaOH, 210, 24h	Absorption 600nm	[197]
$Ce_{1-x}ZrxO_2$	nanoparticles	15	$[NH_4]_2[Ce(NO_3)_6]$,	Raman 461-477	[198]
$La_{0.5}Ba_{0.5}MnO_3$	nanowires	30×10000	$KMnO_4+MnCl_2+La(NO_3)_3+Ba(OH)_2$+KOH, 270, 25h	/	[199]
$La_{0.5}Sr_{0.5}MnO_3$	nanowires	50×1500	$KMnO_4+MnCl_2+La(NO_3)_3+Sr(NO_3)_3$+KOH, 280, 50h	magnetism	[200]

Table 5. The Experimental Conditions and Results of Hybrid Nanostructures

samples	Type of nanostructure	Size (nm)	Starting materials & temperature (°C) & time	Properties	Reference
Ni-Cu	Nanoparticles	12	$NiAc_2$+$CuAc_2$+SDBS+ polyethylene glycol+ammonia, 150,3h	magnetism	[204]
ZnO-Al_2O_3	Nanorods	200×200	$Zn(NO_3)_2$+$Al(NO_3)_3$+ammonia, 20h	/	[205]
C-Fe_3O_4	C-Fe_3O_4 nanoparticle	20-30	$Fe[(NH_2)_2CO]_6(NO_3)_3$+C nanotubes+EN, 200, 50h	Mössbauer	[206]
Ag-C	Nanocables	40×200×10000	Starch+$AgNO_3$, 80, 12h	/	[207]
TiO_2-Fe	Nanotube-nanoparticles	20 in 5×400	$Fe(NO_3)_3$+titanium tetra-isopropoxide+1,4-butanediol, 300,1h	photocatalytic	[208]
TiO_2-Fe_xO_y	Nanoparticles	50	$Fe(NO_3)_3$+titanium tetra-isopropoxide+isopropylalcohol, 200, 1h	photocatalytic	[209]
TiO_2-VO_x	Nanorods	100×1000	Titanate nanotubes+$V_2O_5.nH_2O$, 200, 48h	/	[210]
TiO_2-SiO_2	Nanoparticles	14-16	$TiOSO_4$+tetraethyl orthosilicate, 200, 24h	Raman 100	[211]
TiO_2-CdS	Nanoparticles	8	$Cd(NO_3)_2$+$(NH_4)_2S$+titanium isopropoxide, 200, 12h	/	[212]
A-Fe_2O_3-In_2O_3	Nanoparticles	20	$Fe(NO_3)_3$+$In(NO_3)_3$+H_2O, 400, 4h	Mössbauer	[213]
ZnS-octylamine	Nanosheets	200	$ZnAc_2$+EDTA+OA+Na_2S, 110, 3h	Absorption 286	[214]
Co(II) complex	Nanocluster	/	NH_4NO_3+$H_3[P(Mo_3O_{10})_4$+ $Co(en)_3Cl_3$+1,10-phenanthroline,170, 240h NH_4NO_3+$H_3[P(Mo_3O_{10})_4$+ $Co(en)_3Cl_3$+1,10-phenanthroline,170, 240h	/	[215]

Table 6. The Experimental Conditions and Results III-V Semiconductor and Other Nanostructures

Samples	Type of nanostructure	Size (nm)	Starting materials & temperature (℃) & time	Properties	References
BN	Nano-plates	40×50	$H_3BO_3+Na_3N+N_2H_4.H_2O+NH_4Cl$, 400, 48h	/	[243]
	Nanoparticles nanorods	15 (cBN) 10×200 (oBN)	$H_3BO_3+(CH_3)_3N+N_2H_4.H_2O$, 300, 24h		[244]
GaN			$Ga(C_6H_5N_2O_2)_3+NH[Si(CH_3)_3]_2$,250,2h A		[245]
	nanoparticles	4	2umol/4ml	PL 280	
		12	B 0.4mmol/4ml	PL 315	
		2.5	C 4umol/4ml+4umol+1umolCTAB	PL 265	
		2.5	D 0.568mmol GaCl3/5ml $NH[Si(CH_3)_3]_2$+0.284mmol CTAB	PL265	
	Nanoparticles & nanorods	50 50×300	$GaCl_3+Na_3N$, 260, 48h, annealing at 1000 and solvent wash	PL 320, PL 360	[246]
GaP	nanoparticles	aggregate	$GaCl_3+Na_3P$, 300, 0h	Absorption 400	[247]
		5	$GaCl_3+Na_3P$, 300, 3h	Absorption 470	
		8	$GaCl_3+Na_3P$, 300, 7h	Absorption 490	
		46	$GaCl_3+Na_3P$, 300, 12h	Absorption 510	
InP	nanoparticles	30	$InCl_3+P_4$+ ammonia, 140, 8h	Absorption 910	[248]
	Nanoparticles & nanorods	8, 7×20	$InCl_3+P_4+C_{17}H_{35}COOK$+ ammonia, 140, 8h	Absorption 685	
GaAs	nanoparticles	35	$Ga+As_2O_3+HCl$, 180, 24h	/	[249]
InAs	nanoparticles	30-50	$InCl_3+Zn+As_2O_3$, 120, 18h	Absorption 276 Raman217, 237	[250]
	Nanoparticles	15	$InCl_3+AsCl_3+Zn$, 150, 48h	/	[251-252]
InSb	Nano-cubes	20-60	$InCl_3$+In+benzene, 300, 12h	/	[253]
NiSb	Nanoparticles Nano-dentrite	15	$NiCl_2$+Sb+KBH4+EN, 140, 24h $NiCl_2$+Sb+KBH4+DEN, 140, 24h	Absorption 385 PL 400	[254]

Table 6. The Experimental Conditions and Results III-V Semiconductor and Other Nanostructures (continued)

Samples	Type of nanostructure	Size (nm)	Starting materials & temperature (℃) & time	Properties	References
Ni_2P	nanoparticles	200	$NiCl_2$+P(red)+polyacrylamide,120,20h	PL 346	[255]
		40	$NiCl_2$+P(red)+polyacrylamide,150,20h		
		20	$NiCl_2$+P(red)+polyacrylamide,180,20h		
BaF_2	Nano-whiskers	40×30000	$BaCl_2$+HF+CTAB+cyclohexane+ n-pentanol, 120, 12h	/	[256]
CoPt	nanowires	30×10000	$Pt(CH_3COHCOCH_3)_2$+EN+ $Co_2(CO)_8$, 160, 12h	Magnetism	[257]
C_3N_4	nanoparticles	11	CCl_4+NH_4Cl, 400, 20h	/	[258]
TiB_2	TiB2	15-40	Na+B+$TiCl_4$+benzene, 400, 8h	/	[259]

5.2. Effect of Experimental Conditions on Nanostructures

The experimental conditions refer to starting materials including the starting reactant materials, solvents, surfactants, reaction temperature, and reactive time. They have different effects on the formation, structures, and properties of nanocrystals. The experimental conditions and the experimental results are listed in Table 4.

5.2.1. Effect of Starting Materials on Nanostructures

The effect of starting materials on the nanostructure include the effect of the source of reactants, the effect of the ratios of different reactants, the effect of the amount of the reactants, the effect of solvent and its amount, and the effect of surfactants. The effect of reactant source on nanostructures of ternary and multi-elemental compounds is different and may affect the formation of these compounds. For the formation of $CuFeS_2$ nanorods in the system of CuCl, $FeCl_3$ and $(NH_4)_2S$ [189], $(NH_4)_2S$ and NH_2CSNH_2 are the suitable sulfur source. When elemental S was used instead of $(NH_4)_2S$ keeping other conditions identical, the shape of formed $CuFeS_2$ is irregular nanoparticles and disk-like particles. When Na_2S was used instead of $(NH_4)_2S$, no $CuFeS_2$ could form but Cu_2S and FeS formed. The case of the formation of CuInS2 and $CuSIn_2S_4$ and $PbSnS_3$ is similar [190-191, 194] (Table 4). When CS_2 was replaced by S powder or $CuCl_2$ was replaced by CuCl, keeping other conditions identical, no pure $CuInS_2$ formed. $CuIn_2S_4$ nanorods were formed at 180°C for 6h in the reactive system of CdS nanorods, $InCl_3$, and thiourea. When Na_2S and $(NH_4)_2S$ were used as a sulfur source instead of thiourea, no $CuIn_2S_4$ was obtained. $PbSnS_3$ nanorods formed from the reaction of $PbCl_2$, $SnCl_4$, NH_2CSNH_2, and I_2 [194]. When Na_2S and $(NH_4)_2S$ were used instead of NH_2CSNH_2, keeping other conditions identical, the poorly crystalline $PbSnS_3$ and only PbS were obtained respectively.

The pH has an obvious effect on the formation of ternary compounds. In the formation of $CdWO_4$ nanorods [192], when pH=3-8 keeping other experimental parameters unchanged, well-crystallized $CdWO_4$ nanorods were obtained. When pH>8, the impure $CdWO_4$ formed. For the preparation of nano-sized $CoAl_2O_4$ pigment, $CoAl_2O_4$ nanoparticles could be obtained when 3M NaOH was used to adjust pH. However, when ammonia (pH<14) was used to replace NaOH [197], no $CoAl_2O_4$ formed.

The ratios of reactants have important influence on the formation of ternary compound nanostructure. For the formation of Ag_3CuS_2 nanorods, the relative quantities of AgCl and CuCl played an important role in the purity of the Ag_3CuS_2 sample. For obtaining the pure Ag_3CuS_2, the excessive AgCl is needed. Otherwise formed Cu_2S is difficult to remove [188]. In the formation of $Ce_{1-x}Zr_xO_2$, when the ratio of Ce/Zr is 1/0, 4/1, 1/1, the lattice of $Ce_{1-x}Zr_xO_2$ is cubic and the size is 7nm, 7nm, and 5nm, respectively. When the ratio of Ce/Zr is 1/4, 1/9, 0/1, the lattice of obtained product is tetragonal and size of nanoparticles is 4nm, 6nm, and 6nm, respectively.

The surfactants and a small amount of addictives may affect the size and formation of ternary and multi-elemental compounds. Surfactant Tween-80 contributed to the narrow particle size distribution of $BaTiO_3$ nanocrystals and limitation of further growth and agglomeration [193]. In the preparation of $Ca_{10}(PO_4)_6(OH)_2$, three types of surfactants CTAB, SDS, and PVA, were used. CTAB and SDS-involved samples transformed to nanorods. But PVA-involved sample only gave aggregates [196]. In the system of $PbCl_2$, $SnCl_4$, and NH_2CSNH_2, addition of addictive I_2 could speed the formation of $PbSnS_3$ nanorods [194].

5.2.2. Effect of Reaction Temperature and Reaction Time on Nanostructures

For the formation of ternary and multi-elemental compounds, relatively high temperature and long reaction time are in favor with the formation of 1D nanostructure. For example, $CuFeS_2$ nanorods formed at 200°C and the size of these nanorods increase with the increasing of temperature. But $CuFeS_2$ disk-like nanostructures were obtained at 180℃ [189]. In the formation of $Ca_{10}(PO_4)_6(OH)_2$ nanorods, it was found that 2h and 5h sample both contain fibers, rods and aggregates, but 10h and 16h samples were uniform nanorods with the same size. The effect of reaction time on the various materials is different. For the formation of $CdIn_2S_4$ nanorods, the reaction time is at least 6h when reaction temperature is 180°C. However, when the reaction time is between 6h and 30h, there is no significant effect on $CdIn_2S_4$ nanorods [191]. The case is similar to the effect of reaction time on the Ag_3CuS_2 nanorods [188]. When reaction time varied between 8 and 24h at 150°C, there is no obvious change on the crystallinity, shape and size of Ag_3CuS_2 nanorods. For the formation of SbSI nanorods [195], when the reaction time is 2h, 4h, and 8h while reaction temperature is 180°C, the products in the first two cases is impurity SbSI. The product in third case is SbSI nanorods. When reaction temperature was raised to 220°C, the obtained product is impure SbSI when the reaction time was 1.5h but the product became pure SbSI nanorods when reaction time was 6h.

5.2.3 Effect of Pre-Treatment on the Nanostructures

Pre-treatment has an important effect on the formation of some ternary nanostructure. The preparation of homogeneous aqueous solution of AgCl-Tu (thiourea) and Cu-Tu was a key factor in the formation of Ag_3CuS_2 nanorods [188]. When AgCl, CuCl, thiourea, and water were put into autoclave directly, keeping the other reaction conditions constant, it was found that only small amount of Ag_3CuS_2 formed and product mainly consisted of Cu_2S.

5.2.4. Effect of Shape and Size of Nanostructures on the Properties

The properties of ternary compounds studied here are Raman spectra, photoluminescence, and UV-Vis absorption. Usually, the Raman peaks become weaker and broader with the decreasing of nanocrystal size. The Raman of the $CdIn_2S_4$ nanorods is quite different from that of $CdIn_2S_4$ bulk single crystals. Some Raman peaks disappeared and some became weaker and broader [191]. The reason may originate form the small grain sizes. The Raman of 60nm $PbSnS_3$ nanorods have peaks at 625cm^{-1}, 442 cm^{-1}, 309 cm^{-1}, 201 cm^{-1}, and 143 cm^{-1} but 143 cm^{-1} and 625 cm^{-1} peaks in the Raman of 10-15nm $PbSnS_3$ nanorods disappeared.

Like other nano-materials, the PL peak of ternary compounds may shift to low wavelength with the decreasing of the size. The PL intensity of $CdWO_4$ nanorods increased with the size of nanorods and the peak is at 486nm which is red shifted by 25nm compared to the PL emission of single crystal $CdWO_4$.

Non-uniform size may cause the red-shift of UV-Vis absorption peak. An example is the absorption of $CoAl_2O_4$. 70nm $CoAl_2O_4$ nanoparticles have minimum absorption at 500nm and the maximum at 600nm which is consistent with 100nm $CoAl_2O_4$ nanoparticles but is different from the absorption of 50-200nm $CoAl_2O_4$ [197], which has a red-shift at minimum absorption.

6. Hydrothermal/Solvothermal Preparation of Nanostructures of Hybrid Nanostructure

6.1. Introduction

The hybrid nanostructures prepared through hydrothermal/solvothermal process include Ni-Cu nanoparticles, $ZnO-Al_2O_3$ nanorods, $C-Fe_3O_4$ nanoparticles, Ag-C nanocables, TiO_2-Fe nanotube-nanoparticles, $TiO_2-Fe_xO_y$ nanoparticles, TiO_2-SiO_2 nanorods, TiO_2-CdS nanoparticles, $\alpha-Fe_2O_3-In_2O_3$ nanoparticles, ZnS-octylamine, and Co(II) organic complex nanocluster.

Nanoscale hybrid materials have shown novel optical, catalytic, electrical and magnetic properties with respect to those of pure material and corresponding bulk materials. For instance, it has been reported that the noble-transition-metal alloys such as CuNi, AgPd and AuPt as atomic-sized metallic point contacts may be applied in small-scale electronic devices [201]. Ni-Co alloys have some applications in hydrogen evolution as coating electrodes, catalysts for N_2O decomposition ans organic synthses [202]. Fe-rich Au-Fe alloy films possess magnetoptical and optical properties [203]. The hybrid compounds studied here also have improved properties. Ni-Cu alloy has magnetic application [204]. Fe-TiO_2 and Fe_xO_y-TiO_2 nanoparticles have high photo-catalytic activity compared to each component [208-209]. TiO_2-SiO_2 nanocomposite has high phase stability [211]. Utilization of nanosized CdS to couple with nanocryslline TiO_2 could improve the efficiency of photocatalysis [212]. $In_2O_3-\alpha-Fe_2O_3$ has improved sensing properties [213]. ZnS-octylamine nanosheets exhibit a great blue shift and stronger intensity compared to the bulk ZnS [214].

6.2. Effect of Experimental Conditions on Nanostructures

The experimental conditions of preparing hybrid nanostructures mainly include the surfactants, pH, reaction temperature, and reaction time. Surfactants and pH may affect the formation of hybrid compounds, the shape and size of the hybrid compounds. In most cases, relatively high reaction temperature and long reaction time lead to the large size. The experimental conditions and experimental results of these nano-structural compounds are listed in Table 5.

6.2.1. Effect of Starting Materials on Nanostructures

In the formation of hybrid nanostructures, the effect of the surfactants and pH was studied in most of cases. Surfactants may affect the formation of hybrid nanostructures. An example is the preparation of Ni-Cu alloy nanoparticles [204]. When polymer polyethylene glycol octyphenyl ether was employed, the alloy formed. Elemental Ni and Cu obtained in absence of polyethylene glycol while keeping other conditions identical. Another case is a surfactant could affect the shape of hybrid nanostructures. For instance, EDTA (ethylene diamine tetraacetic acid) and OA (octylamine) plays an important role in the control over the morphology of ZnS-octylamine [214]. In absence of EDTA and OA, the shape of the prepared product is spherical. In presence of EDTA and OA, the morphology of obtained product is nanosheets.

Also pH has different effect on the size of hybrid nanostructures. In the preparation of Zn-Al alloy nanorods [205], pH has an obvious effect on the size of Zn-Al nanorods. The size of the nanorods prepared at pH=8 is much larger than that prepared at pH=5 and pH=10 while keeping other experimental conditions identical.

6.2.2. Effect of Reaction Temperature and Reaction Time on Nanostructures

Generally, relative high reaction temperature and long reaction time lead to large size. For example, the size of ZnS-octylamine hybrid nanosheets increased with the reaction temperature [214]. As for Ag-C core-sheath nanorods [207], when reaction temperature is 80°C, the diameter of Ag-core nanorods is 40nm. When reaction temperature is 200°C, the diameter of Ag-core nanorods could increase to 80-150nm. However, for Al_2O_3-ZnO nano-composite nanorods prepared from $Al(NO_3)_3$ and $Zn(NO_3)_2$ under hydrothermal conditions [205], higher reaction temperature did not change the diameter but shortened the length of the nanorods. When the reaction temperature is 180°C, the average size of Al_2O_3-ZnO nanorods is 20×200-400nm. When the reaction temperature is 200°C, the average size of Al_2O_3-ZnO nanorods is 20×150-200nm.

6.2.3. Effect of Hybrid Compound and the Ratios of Their Components on the Properties

Hybrid compounds usually have different properties compared to the every component and the properties may change with the ratios of the components. Like other nanostructures, the different shapes and sizes have a different effect on the properties as well. For instance, the saturation magnetization Ms of Ni-Cu alloy ($Ni_{0.63}Cu_{0.47}$) is 18emug-1, which is much lower than the value of 34.6emug-1 of bimetallic nanocrystalline ($Ni_{0.63}Cu_{0.47}$) [204]. The electrical conductivity of nano-composites may be better than corresponding component. The electrical conductivity of C-Fe_3O_4 nanocomposite is 2.5Scm-1 while that of C and Fe_3O_4 is 10^2-$10^{-4}$$Scm^{-1}$ and 0.236Scm^{-1} [206], respectively.

The optical properties of nano-composite could be different from corresponding components. First, the absorption spectrum of nano-composite is different from the corresponding component. For example, Fe-TiO_2 nano-composite has an absorption at 370nm but the absorption of pure TiO_2 is 350nm [208]. The absorption peaks (248nm and 286nm) of ZnS-Octylamine shifted to short wavelength as well compared to bulk ZnS (peak at 344nm) [214]. Second, PL spectra of nano-composite may be different from the corresponding component. The PL intensity of Fe_xO_y-TiO_2 (10% Fe) nano-composite synthesized by hydrothermal method is stronger than pure TiO_2 nanoparticles. Third, the Raman could be changed in the nano-composite compare to the component. The Raman intensity TiO_2-SiO_2 nanocomposite increased with the decreasing of SiO_2.

The photocatalyst ability of nano-composite is also different from the corresponding component. The speed of methanol decomposition is slightly decreased when Fe-TiO_2 was used as catalyst instead of pure TiO_2 at UV light [208]. However, the decomposition degree of $CHCl_3$ was nearly the same by Fe_xO_y-TiO_2 ($Fe(NO_3)_3$ 10%) nano-composite prepared through hydrothermal route as catalyst and by the pure TiO_2 as catalyst [209]. Changing the percentage of Fe%, it was found that $CHCl_3$ decomposition was enhanced by up to 10wt% of Fe concentration and then decreased with an increase in the Fe concentration [208] at UV 365nm. In the case of CdS-TiO_2 nano-composite particles [212], the nano-sized CdS-TiO_2

showed high visible-light activity on the methylene blue reduction and the catalyst ability increased with the increasing of concentration of CdS.

7. Hydrothermal/Solvothermal Preparation of Nanostructures of III-V Semiconductor and other Nanostructure

7.1. Introduction

III-V semiconductor and other nanostructures discussed here are BN nano-plates, nanoparticles and nanorods, GaN nanoparticles and nanorods, GaP nanoparticles, InP nanoparticles, GaAs nanoparticles, InSb nanocubes, Ni_2P nanoparticles, BaF_2 nano-whiskers, CoPt nanowires, C_3N_4 nanoparticles and TiB_2 nanoparticles.

These materials have a variety of important properties and applications. Cubic BN is the second hardest material and does not react with ferrous metals. Moreover, cBN is more stable than diamond when they are in the air at high temperature. In addition, BN possesses excellent physical and chemical properties, such as high thermal conductivity, high electrical resistivity, optical transparency over a wide spectral range [216-217]. It can be used in the fabrication of cutting tools, super-hard protective coating and grinding materials [216, 218-223], microwave devices, short wavelength optoelectronic devices, high temperature and high power semiconductor devices [224-226]. Another nitride GaN (Eg=3.4eV) has useful band gaps in the UV-visible region and form a wide range of solid solutions [246]. Phosphide GaP is an indirect-gap semiconductor and has many important uses in microelectronic devices [247]. Arsenide GaAs is one of the most important direct band-gap semiconductors and has high electron mobility and an exciton diameter of 19nm, which has been widely used for laser diodes, full-color flat-panel displays, high-speed transistors and is more likely to find applications in optical devices [227]. Another arsenide InAs is one of the narrow band-gap semiconductors (Eg=0.35eV) and has a rather great excitonic diameter of 62nm [228] which is expected to possess pronounced quantum confinement effects due to the relative covalent bonding and direct band gap structure [229] and depending on the degree of size quantization. Its optical band-gap is potentially tunable from the near-infraed to the visible range. Antimonide InSb has a small bandgap (0.17eV) and the largest room temperature carrier mobility, which makes it have potential for making high speed transistors [230-232]. Other nanostructures such Ni_2P, BaF_2, CoPt, C_3N_4, and TiB_2 discussed in this section also have interesting properties and potential applications. Nickel phosphides have found application in wear-proof materials [233]. Due to the special structure of dinichel phosphide nanocrystals, they are promising candidates for luminescent devices, magnetic or electronic components [234]. BaF_2 is one of the dielectric fluorides that have a wide range of potential applications in microelectronic and optoelectronic devices, such as wide-gap insulating over-layers, gate dielectrics, and more advances three-dimensional structure devices [235]. BaF_2 activated with rare-earth ions has also been reported to display unique luminescence properties and can be used as scintillators [236]. C_3N_4 may have a hardness exceeding that of diamond [237-238] and it also has high wear resistance and low coefficient of friction [255]. Titanium diboride is an important material in diecasting and dry cutting industry for ware, corrosion and erosion protection [240-241]. Titanium di-boride is a promising interconnection material for intergrated circuits for its high temperature stability and good electrical conductivity [242].

7.2. Effect of Experimental Conditions on Nanostructures

Most of III-V compounds and other compounds prepared thorough the hydrothermal/solvothermal route are nanoparticles. The main discussion on them is about the effect of the experimental conditions on the type of crystals and the size of nanocrystals. The experimental conditions and results are listed in Table 6.

7.2.1. Effect of Starting Materials on Nanostructures

The starting materials, such as reactants, solvents, surfactants, and addictives, may have different effects on the structures of nanocryastals. They could affect the type and crystalline quality of nanocrystals. In the formation of BN nano-plates, the addition of $N_2H_4.H_2O$ and NH_4Cl played an important role. The as prepared BN had a rather poor crystalline quality and the crystal type is hBN in absence of $NH_2NH_2.H_2O$. When $NH_2NH_2.H_2O$ was added to the reaction system, keeping other conditions identical, the crystalline quality was improved evidently and the XRD peaks indicated they are cBN and oBN. On the other hand, the yield (20%) of product with $NH_2NH_2.H_2O$ was four times as that (5%) without $NH_2NH_2.H_2O$. When another additive NH_4Cl was added to the reaction system containing $NH_2NH_2.H_2O$, the yield of the product increased to 65%, which is more than three times as that (20%) without NH_4Cl. Also the relative content of cBN increased.

The starting materials may influence the size of as prepared nanoparticles. One could see from Table 6 that GaN nanoparticles [245] have different sizes of 4nm, 12nm, 2.5nm and 2.5nm when the corresponding starting materials is 2umol $Ga(C_6H_5N_2O_2)_3$ + 4ml $NH[Si(CH_3)_3]_2$, 0.4mmol $Ga(C_6H_5N_2O_2)_3$+4ml $NH[Si(CH_3)_3]_2$, 4umol$Ga(C_6H_5N_2O_2)_3$+4ml $NH[Si(CH_3)_3]_2$+1umol CTAB, 0.568mmol $Ga(C_6H_5N_2O_2)_3$+5ml $NH[Si(CH_3)_3]_2$+0.284mmol CTAB. The starting materials could affect the size and shape of prepared nanostructures simultaneously. It could seen from Table 6 that InP nanoparticles with average size of 30nm were obtained when the starting materials were $InCl_3$+P_4+ammonia while 8nm of nanoparticles and 7×20nm of InP nanorods were obtained when the starting materials were $InCl_3$+P_4+ammonia+ $C_{17}H_{35}COOK$ keeping other conditions identical. The starting material solvent can also change the shape of III-V nanostructures. In the reaction system of $NiCl_2$+Sb+KBH_4, when EN (ethylenediamine) was used as solvent, the prepared NiSb is nanoparticles; when DEN (diethylamine) was employed instead of EN keeping other conditions identical, the obtained NiSb is nano-dentrites.

7.2.2. Effect of Reaction Temperature and Reaction Time on Nanostructures

As we discussed in other sections, hydrothermal /solvothermal reaction temperature and reaction time have different effects on the nanostructures of elements, oxides and hydroxides, chalcogenides, ternary and multi-elemental compounds, hybrid and composite compounds. The reaction temperature and reaction time have a similar but different effect on the III-V compound nanostructures as well. Generally, the relatively high reaction temperature and long reaction time result in the good crystallization and larger size. For GaP nanoparticles formed in the reaction system of $GaCl_3$+Na_3P, when reaction time is 0h, the obtained GaP nanoparticles aggregated with poor crystallization and the size can not be measured. When the reaction time is 3h, 7h, and 12h, the corresponding size of obtained GaP nanoparticles with good crystallization is 5nm, 8nm, 46nm. However, there is some exception. One could see from Table 6 that the size of Ni_2P nanoparticles decreased with the increasing of reaction temperature. When the reaction temperature is 120°C, 150°C, and 180°C, the corresponding size of Ni_2P nanoparticles is 200nm, 40nm, and 20nm while keeping other experimental conditions identical.

7.2.3. Effect of Post-Treatment on the Nanostructures

The post treatment of hydrothermal/solvethermal reaction could affect the size and structure of the nano-materials. For GaN nanoparticles prepared in the reaction system of $GaCl_3$+Na_3N+toluene [246], they are 50nm of nanoparticles and some rodlike structures embed in a salt and nanoparticles matrix before post-treatment of hydrothermal reaction. However, they became nanorods with an average size of 100nm×2um after glycerol-ethanol washing and nanorods with an average size of 50×300nm after annealing at 1000°C and washing by water. GaN nanoparticles prepared in the reaction system of $GaCl_3$+Na_3N+THF [246] have a similar size and shape as in toluene. However, the washed material by glycerol/ethanol exhibits a more highly interconnected network of nanoparticles. Also crystallization of CoPt alloy nanowires prepared from $GaCl_3$+Na_3N, was improved and the length of CoPt became much longer after annealing at 580°C [258].

7.2.4. Effect of Shape and Size of Nanostructures on the Properties

The size and the shape of nanostructures have an extinct influence on the different properties of III-V nano-materials. The photoluminescence and absorption shift to low wavelength with the decreasing of the size when the size is smaller than the Bohr radius. One could see from Table 6 that the corresponding PL peak of 2.5nm, 4nm, and 12nm of GaN nanoparticles is at 265nm, 280nm, 315nm. The PL peak [246] of GaN nanorods with an average diameter of 100nm×2um from $GaCl_3$+Na_3N+toluene and glycerol-ethanol washing is at 320nm. However, the PL peak of GaN with an interconnected network of nanoparticles prepared from $GaCl_3$+Na_3N+toluene and under same post-treatment is at 360nm. The PL peak does not shift with the change of size of nanoparticles when the emission was caused by disfigurement. An example is Ni_2P nanoparticles [255]. The PL peak is the same and is at 346nm for 200nm Ni_2P particles, 40nm Ni_2P nanoparticle, and 20nm Ni_2P nanoparticles.

The absorption spectrum of nanostructure could shift with the size of the nano-materials. The absorption of aggregated GaP (A), 5nm GaP nanoparticles (B), 8nm GaP nanoparticles (C), and 46nm GaP nanoparticles (D) is quite different [247]. The absorption spectrum of A is only an inconspicuous peak at about 400nm, indicating the amorphous nature of A. The absorption of spectrum of B and C shows a broad peak at 470nm and 490nm, respectively. The absorption spectrum of D has an edge peak at 510nm. The absorption edge shifts to the red due to the growth of grains. For InP with different size and shape, the absorption is different as well [248]. InP nanoparticles with an average size of 30nm have an absorption at 910nm while InP nanoparticles-nanorods with average size 8nm and 7×20nm have an absorption at 685nm.

Conclusion

We have reviewed the hydrothermal/solvethermal route to nano-structural materials. These materials include elements, oxides and hydroxides, chalcogenides, ternary and multi-elemental compounds, hybrid and composite nano-compounds, III-V and other compounds. The nanostructures include three-dimensional (3D) nanostructures, such as nanoparticles, nanospheres, nanocells, nanocubes, nano-spindles, nano-dentrites, and flower-like nanostructure, 2D nanostructures, such as nanosheets, nanoplates, nanoribbons, 1D nanostructure such as nanofibers, nanorods, nanowires, nanobelts, nanotubes and porous nanotubes, and 0D nanostructure nanodots. The effect of starting materials, reaction temperature, reaction time, pre-treatment and post-treatment on the size and shape were discussed. The effect of size and shape on the properties of nanostructures was discussed as well.

Reference

[1] Byrappa, K.; Masahiro Yoshimura *Handbook of Hydrothermal Technology*, Park Ridge, NJ, Noyes Publications; Norwich, NY, William Andrew Publications, c2001.

[2] Michael Rajamathi, Ram Seshadri *Curr. Opin. Solid Mater. Sci.*, 2002, 6, 337-345.

[3] Ujjal K Gautam; Monmita Ghosh; Michael Rajamathi; Ram Seshadri *Pure Appl. Chem.* 2002, 74, 1643-1649.

[4] Schimd, D.; *Clusters and Colloids* VCH Press, NY, 1994: Furstner, A., VCH Weinheim, NY, 1996.

[5] Novak, J. P.; Brousseau, F. W.; Vance, R. C.; Johnson, B. J.; Lemon, J. T.; Hupp, D. I.; Feldheim, J. *J. Am. Chem. Soc.* 2000, 122, 12029-12030 ; Kim, J. Y.; Yoon, S. B.; Yu, J.-S. *Chem. Commun.* 2003, 790-791.

[6] Nicewarner-Pina, S. R.; Freeman, R. G.; Reiss, B. D.; He, L.; Pena, D. J. *Science* 2003, 294, 137-141.

[7] Frank, S.; Ponchareal, S.; Wang, Z. L.; Deheer, W. A. *Science* 1998, 280, 1744-1746.

[8] Dillon, A. C.; Jones, K. M.; Bekkedahl, T. A.; Kiang, C. H.; Bethune, D. S.; Heben, M. J. *Nature* 1997, 386, 377-379.

[9] Fan, S.; Chapline, M. G.; Frank, N. R.; Tommbler, T. W.; Cassell, A. M.; Dai, H. J. *Science* 1999, 283, 512-514.

[10] Kim P. Lieber, C. M. *Science* 1999, 286, 2148-50.

[11] Rosemary, M. J.; Pradeep, T. J. *Collo. Inter. Sci.* 2003, 268, 81-84.

[12] Wei, G. D.; Nan, C. W.; Deng, Y.; Lin, Y. H. *Chem. Mater.* 2003, 15, 4436-41.

[13] Liu, Z. P.; Yang, Y.; Liang, J. B.; Hu, Z. K.; Li, S.; Peng, S.; Qian, Y. T. *J. Phys. Chem. B* 2003, 107, 12658-661.

[14] Wang, D. B.; Yu, D. B.; Peng, Y. Y.; Meng, Z. Y.; Zhang, S. Y.; Qian, Y. T. *Nanotechnology* 2003, 14, 748-751.

[15] Hu, H. M.; Yang, B. J.; Li, Q. W.; Liu, X. Y.; Yu, W. C.; Qian, Y. T. *J. Cryst. Growth* 2004, 261, 485-89.

[16] Gao, Y. H.; Niu, H. L.; Zeng, C.; Chen, Q. W. *Chem. Phys. Lett.* 2003, 367, 141-144.

[17] Liu, X. Y.; Zeng, J. H.; Zhang, S. Y.; Zheng, R. B.; Liu, X. M.; Qian, Y. T. *Chem. Phys. Lett.* 2003, 374, 348-52.
[18] Zhang, Y. X.; Li, G. H.; Zhang, L. D. *Inorg. Chem. Comm.* 2004, 7, 344-346.
[19] Hou, Y. L.; Gao, S. J. *Alloys Comp.* 2004, 365, 112-116.
[20] Liu, Z. P.; Li, S.; Yang, Y.; Peng, S.; Hu, Z. K.; Qian, Y. T. *Adver. Mater.* 2003, 15, 1946-948.
[21] Gao, S.; Zhang, J.; Zhu, Y. F.; Chen, C. M. *New J. Chem.* 2004, 24, 739-740.
[22] Wang, Z. H.; Chen, X. Y.; Liu, J. W.; Yang, X. G.; Qian, T. Y. *Inorg. Chem. Comm.* 2003, 6, 1329-331.
[23] Xu, L. Q.; Ding, Y. W.; Xi, G. C.; Zhang, W. Q.; Peng, Y. Y.; Yu, W. C.; Qian, Y. T. *Chem. Lett.* 2004, 33, 592-593.
[24] Wei, G. D.; Deng, Y.; Lin, Y. H.; Nan, C. W. *Chem. Phys. Lett.* 2003, 372, 590-594.
[25] Mo, M. S.; Zeng, J. H.; Liu, X. M.; Yu, W. C.;, Zhang, S. Y.; Qian, Y. T. *Adver. Mater.* 2002, 14, 1658-62.
[26] Zhang, W.; Ma, D. K.; Liu, J. W.; Kong, L. F.; Yu, W. C.; Qian, Y. T. *Carbon* 2004, 42, 2329-66.
[27] Jiang, Y.; Wu. Y.; Zhang, S. Y.; Xu, C. Y.; Yu, W. C.; Xie, Y.; Qian, Y. T. *J. Am Chem. Soc.* 2000, 122, 12383-384.
[28] Kuang, Q.; Xie, S. Y.; Jiang, Z.Y.; Zhang, X. H.; Xie, Z. X.; Huang, R. B.; Zheng, L. S. *Carbon* 2004, 42, 1737-741.
[29] Hu, G.; Cheng, M.; Ma, D.; Bao, X. *Chem. Mater.* 2003, 15, 1470-1473
[30] Ye, H.; Naguib, N.; Gogotsi, Y.; Yazicioglu, A. G.; Megaridis, C. M. *Nanotechnology* 2004, 15, 232-236.
[31] Moreno, J. M. C.; Yoshimura, M. *J. Am. Chem. Soc.* 2001, 123, 741-742.
[32] Moreno, J. M. C.; Fujino, T.; Yoshimura, M. *Carbon* 2001, 39,615-628.
[33] Moreno, J. M. C.; Swamy, S. S.; Fujino, T.; Yoshimura, M. *Chem. Phys. Let.* 2000, 329, 317-322.
[34] Suchanek, W. L.; Libera, J. A.; Gogotsi, Y.; Yoshimura, M. *J. Solid State Chem.* 2001, 160, 184-188.
[35] Gogotsi, Y.; Libera, J. A.; Yoshimura, M. *J. Mater. Res.* 2000, 15, 2591-2594
[36] Li, Y. D.; Sui, M.; Ding, Y.; Zhang, G. H.; Zhuang, J.; Wang, C. *Adv. Mater.* 2000, 12, 818-821.
[37] Teeramongkonrasmee, A.; Sriyudthsak, M. *Sensor Actuator* 2000, B66, 256-259.
[38] Ansari, S. G.; Boroojerdian, P.; Sainker, S. R.; Karekar, R. N.; Aiyer, R. C.; Kulkarni, S. K. *Thin Solid Films* 1997, 295, 271-276.
[39] Ferrere, S.; Zaban, A.; Gsegg, B. A. *J. Phys. Chem. B* 1997, 101, 4490-93.
[40] Rowlette, J. J., Attia, H. I. *Proc. Electrochem. Soc.* 1987, 87, 209-219.
[41] Stampfl, S. R.; Chen, Y.; Dumesis, J. A.; Chen, N.; Hill, J. *J. Catalyst* 1987, 105, 445-454.
[42] Boudeville, Y.; Figuras, F.; Forissier, J.; Portefaix, J. L.; Vedrine, J. C. *J. Catal.* 1979, 58, 52-60.
[43] Pillep, B.; Behrens, P.; Schubert, U. A.; Spengler, J.; Knozinger, H. *J. Phys. Chem. B* 1999, 103, 9595-9603.
[44] Ozawa, K.; Sakka, Y.; Amamo, A. *J. Mater. Res.* 1998, 13, 830-833.
[45] Sravani, C.; Reddy, K. T. R.; Hussain, O. M.; Reddy P. J. *Thin Solid Films* 1994, 253, 339-343.

[46] Su, L. M.; Grote, N.; Schmitt, F. *Electron Lett.* 1984, 20, 716-717.
[47] Benko, F. A.; Koffyberg, F. P. *Solid State Commun.* 1986, 57, 901-903.
[48] Lide, D. R. *CRC Handbook of Chemistry and Physics* 77th edn., CRC press, Boca Raton, 1996/1997, 3-278, 12-97.
[49] Fleet, M. E. *J. Solid State Chem.* 1986, 62, 75-82.
[50] Hagfeldt, A.; Gratzel, M. *Chem. Rev.* 1995, 95, 49-68.
[51] Tian, Z. R; Tong, W.; Wang, J. Y.; Duan, N. G.; Krishnan, V.V.; Suib, S. L. *Science* 1997, 276, 926-930
[52] Matsuda, S.; Kato A. *Appl. Catal.* 1983, 8, 149-165.
[53] Yeh, Y. C.; Tseng, T. T.; Chang, D. A. *J. Am. Ceram. Soc.* 1990, 73, 1992-98.
[54] Centi, G. *Catal. Today* 1993, 16, 5-26.
[55] Yamashita, T.; Vannice, A. *J. Catal.* 1996, 163, 158-168.
[56] Foss, C. A.; Hornyak, G. L.; Stockert, J. A.; Martin, C. R. *J. Phys. Chem.* 1992, 96, 7497-99.
[57] Routkevitch, D.; Bigioni, T.; Moskovits, M.; Xu, J. M. *J. Phys. Chem.* 1996, 100, 14037-047.
[58] Davydov, D. N.; Haruyama, J.; Routkevitch, D.; Statt, B. W.; Ellis, D.; Moskovits, M.; Xu, J. M. *Phys. Rev. B: Condens. Matter.* 1998, 57, 13550-553.
[59] Schlottig, F.; Textor, M.; Spencer, N. D.; Sekinger, K.; Shnaut, U.; Paulet, J. F.; Fresenius J. *Anal. Chem.* 1998, 361, 684-686.
[60] Fan, W.; Sun, S.; Song, X.; Zhang, W.; Yu, H.; Tan, X.; Cao, G. *J. Solid State Chem.* 2004, 177, 2329-338.
[61] Yan, L.; Zhuang, J.; Sun, X.; Deng, Z.; Li, Y. *Mater. Chem. Phys.* 2002, 76, 119-122
[62] Zhu, H.; Wang, Y.; Wang, N.; Li, Y.; Yang, J. *Mater. Lett.* 2004, 58, 2631-2634.
[63] Guo, C.; Cao, M.; Hu, C. *Inorg. Chem. Commun.* 2004, 7, 929-931.
[64] Chen, D.; Gao, L. *Chem. Phys. Lett.* 2004, 398, 201-206.
[65] Chen, D.; Gao, L. *J. Colloid Inter. Sci.* 2004, 279, 137-142.
[66] Fujihara, S.; Maeda, T.; Ohgi, H.; Hosono, E.; Imai, H.; Kim, S. H. *Langmuir* 2004, 20, 6476-6481.
[67] Torkhov, D. S.; Burukhin, A. A.; Churagulov, B. R.; Rumyanseva, M. N.; Maksimov, V. D. *Inorg. Mater.* 2003, 39, 1158-1162.
[68] Bertone, J. F.; Cizeron, J.; Wahi, R. K.; Bosworth, J. K.; Colvin, V. L. *NanoLetters* 2003, 3, 655-659.
[69] Zhang, Y.; Li, G.; Zhang, L. *Chem. Lett.* 2004, Vol.33, No. 3, 334-335.
[70] Ghosh, M.; Rao, C. N. R. *Chem. Phys. Lett.* 2004, 393, 493-497.
[71] Wang, J.; Gao, L. *Solid State Communi.* 2004, 132, 269-271.
[72] Qiu, M.; Wong, K. S.; Wu, M.; Lin, W.; Xu, H. *Appl. Phys Lett.* 2004, 84, 2739-741.
[73] Cheng, B.; Samulski, E. T. *Chem. Commun.* 2004, 986-987.
[74] Choy, J. H.; Jang, E. S.; Won, J. H.; Chung, J. H.; Jang, D. J.; Kim, Y. W. *Appl. Phys. Lett.* 2005, 84, 287-289.
[75] Zhang, X. Y.; Dai, J. Y.; Wang, N.; Chan, H. L. W.; Choy, C. L. *Chem. Phys. Lett.* 2004, 393, 17-21.
[76] Zhang, H.; Yang, D.; Ma, X.; Ji, Y.; Xu, J.; Que, D. *Nanotechnology* 2004,15,622-626
[77] Chandrappa, G. T.; Steunou, N.; Cassaignon, S.; Bauvais, C.; Livage, J. *Catalysis Today* 2003, 78, 85-89.
[78] Liu, J.; Li, Q.; Wang, T.; Yu, D.; Li, Y. *Angew. Chem. Int. Ed.* 2004, 43, 5048-5052

[79] Kominami, H.; Miyakawa, M.; Murakami, A.; Yasuda, T.; Kohno, M.; Onoue, S.; Kera, Y.; Ohtana, B. *Phys. Chem. Chem. Phys.* 2001, 3, 2697-2703.
[80] Deng, Y.; Wu, J.; Liu, J.; Wei, G.; Nan, C. *J. Phys. Chem. Solids,* 2003, 64, 607-610.
[81] Ma, X.; Zhang, H.; Ji, Y.; Xu, J.; Yang, D. *Mater. Lett.* 2004, 58,1180-1182.
[82] Tang, B.; Ge, J.; Wu, C.; Zhuo, L.; Niu, J.; Chen, Z.; Shi, Z.; Dong, Y. *Nanotechnology* 2004, 15, 1273-1276.
[83] Fang, Y. P.; Xu, A. W.; You, L. P.; Song, R. Q.; Yu, J. C.; Zhang, H. X.; Li, Q.; Liu, H. Q. *Adv. Funct. Mater.* 2003, 13, 955-960.
[84] Lian, S.; Wang, E.; Gao, L.; Kang, Z.; Wu, D.; Lan, Y.; Xu, L. *Solid State Commun.* 2004, 132, 375-378.
[85] Hou, Y.; Yu, J.; Gao, S. *J. Mater. Chem.* 2003, 13, 1983-1987.
[86] Thimmaiah, S.; Rajamathi, M.; Singh, N.; Bera, P.; Meldrum, F.; Chandrasekhar, N.; Seshadri, R. *J. Mater. Chem.* 2001, 11, 3215-3221.
[87] He, T.; Chen, D.; Jiao, X;. Xu, Y.; Gu, Y. *Langmuir* 2004, 20, 8404-8408.
[88] Matsui, K.; Kyotani, T.; Tomita, A. *Adv. Mater.* 2002, 14, 1216-1218.
[89] Yuan, Z. Y.; Ren, T. Z.; Du, G.; Su, G. L. *Chem. Phys. Lett.* 2004, 389, 83-86.
[90] Wang, X.; Li, Y. *J. Am. Chem. Soc.* 2002, 124, 2880-2881.
[91] Wu, C.; Xie, Y.; Wang, D.; Yang, J.; Li, T. *J. Phys. Chem. B* 2003, 107, 13583-13587.
[92] Zhang, W.; Yang, Z.; Liu, Y.; Tang, S.; Han, X.; Chen, M. *J. Cryst. Growth* 2004, 263, 394-399.
[93] Ma, R.; Bando, Y.; Zhang, L.; Sasaki, T. *Adv. Mater.* 2004, 16, 918-922.
[94] Bavykin, D.V.; Parmon, V. N.; Lapkin, A. A.; Walsh, F. C. *J. Mater. Chem.* 2004, 14, 3370-3377.
[95] Tsai, C. C.; Teng, H. *Chem. Mater.* 2004, 16, 4352-4358.
[96] Suzuki, Y.; Yoshikawa, S. *J. Mater. Res.* 2004, 19, 982-985.
[97] Zhang, Y. X.; Li, G. H.; Jin, Y. X.; Zhang, Y.; Zhang, J.; Zhang, L. D. *Chem. Phys Lett.* 2002, 365, 300-304.
[98] Zhang, Q.; Gao, L. *Langmuir* 2003, 19, 967-971.
[99] Reddy, K. M.; Guin, D.; Manorama, S. V.; Reddy, A. R. *J. Mater. Res.* 2004, 19, 2567-575.
[100] Kim, C.S.; Moon, BK, Park, J.H.; Chung, S.T. Son S.M. *J. Cryst. Growth* 2003, 254, 405-410.
[101] Andersson, M.; Osterlund, L.; Ljungstrom, S.; Palmqvist, A. *J. Phys. Chem. B.* 2002, 106, 10674-679.
[102] Wu, M.; Lin, G.; Chen, D.; Wang, G.; He, D.; Feng, S.; Xu, R. *Chem. Mater.* 2002, 14, 1974-1980.
[103] Zhang, D.; Yoshida, T.; Furuta, K.; Minoura, H. *J. Photochem. Photobio. A* 2004, 164, 159-166.
[104] Kolen'ko, Y. V.; Maximov, V. D.; Garshey, A. V.; Meskin, P. E.; Oleynikov, N. N.; Churagulov, B. R. *Chem. Phys. Lett.* 2004, 388, 411-415.
[105] Jeon, S.; Braun, P. V. *Chem. Mater.* 2003, 15, 1256-1263.
[106] Kavan, L.; Kalbac, M.; Zukalova, M.; Exnar, I.; Loreenzen, V.; Nesper, R.; Graetzel, M. *Chem. Mater.* 2004, 16, 477-485.
[107] Bawendi, M. G.; Steigerwald, M. L.; Brus, L. E. *Annul Rev. Phys. Chem.* 1990, 41, 477-496.

[108] Peyghambarian, N.; Fluegel, B.; Hulin, D.; Migus, A.; Joffre, M.; Antonetti, A.; Koch, S. W.; Lindberg, M. *IEEE J. Quantum Electron* 1989, 25, 2516-522.

[109] Loferski, J. J. *J. Appl. Phys*. 1956, 27, 777-784.

[110] Chun, D.; Walser, R. M.; Bene, R. W.; Courtney, T. H. *Appl. Phys. Lett*. 1974, 24, 479-481.

[111] Arivuoli, D.; Gnanam, F. D.; Ramasamy, P. *J. Mater. Sci. Lett*. 1980, 7, 711-713.

[112] Ennaoui, A. H. *Tributsch. Sol. Cells* 1984, 13, 197-200.

[113] Ruizhitzky, E.; Jimenez, R.; Casal, B.; Manriquez, V.; Ana, A. S.; Gonzalez, G. *Adv. Mater*.1993, 5, 738-741.

[114] Miremadi, B. K.; Morrison, S. R. *J. Catal*. 1987, 103, 334-345.

[115] Dejong, A. M.; Debeer, V. H. J.; Vanveen, J. A. R.; Niemantsverdrier, J. W. *J. Phys. Chem*.1996, 100, 17722-724.

[116] Chauss, F. J. *Solid Lubricants and Self-Lubricating Solids*, Acedemic Press, New York, London, 1972.

[117] Chen, W.; Joly, A. G.; Zhang J. Z. *Phys. Rev. B* 2000, 64, 41202

[118] Yang, H. S.; Holloway, P. H.; Ratma, B. B. *J. Appl. Phys*. 2003, 93, 586-592.

[119] Do, Y. R.; Kim, Y. C.; Cho, S. H.; Ahn. J. H.; Lee, J. G. *Appl. Phys. Lett*. 2003, 82, 4172-174.

[120] Xu, C. N.; Watanabe, T.; Akiyana, M.; Zheng, X. G. *Mater. Res. Bull*. 1999, 34, 1491-500.

[121] Chen, W.; Wang, Z. G; Lin, Z. J.; Lin, L. Y. *Appl. Phys. Lett*. 1997, 70, 1465-467.

[122] Bredol, M.; Merichi, J. *J. Mater. Sci*. 1998, 33, 471-476

[123] Calandra, P.; Goffredi, M.; Liveri, V. T. *Colloids Surf. A*. 1999, 160, 9-13.

[124] Prevenslik, T. V. *J. Lumin*. 2000, 87, 1210-212.

[125] Alivisatos, A. P. *Science*1996, 271, 933-937.

[126] Colvin, V. L.; Schlamp, M. C.; Alivisatos, A. P. *Nature* 1994, 370, 354-357.

[127] Klein, D. L.; Roth, R.; Lim, A. K. L.; Alivisatos, A. P. *Nature* 1997, 389, 699-701.

[128] Bruchez Jr., M.; Moronne, M.; Gin, P.; Weiss, S.; Alivisatos, A. P. *Science* 1998, 281, 2013-016.

[129] Rao, M. M.; Jayvalakshmi, M.; Reddy, R. S. *Chem. Lett*. 2004, 33, 1044-1045.

[130] Ji, Y.; Zhang, H.; Ma, X.; Xu, J.; Yang, D. *J. Phys. Condensed Matter* 2003, 15, L661-L665.

[131] Yang, Q.; Tang, K.; Wang, C.; Zhang, D.; Qian, Y. T. *J. Solid State Chem*. 2002, 164, 106-109.

[132] Mo, M. S.; Shao, M. W; Hu, H. M.; Yang, L.; Yu, W. C.; Qian, Y. T. *J. Cryst. Growth* 2002, 244, 364-368.

[133] Yu, D.; Wang, D.; Zhang, S.; Liu, X.; Qian. Y. T. *J. Cryst. Growth* 2003, 249, 195-200.

[134] Wang, D.; Yu, D.; Mo, M.; Liu, X.; Qian, Y. T. *Solid State Commun*. 2003, 125, 475-479.

[135] Xiang, J.; Yu, S.H.; Liu, B.; Xu, Y.; Gen, X.; Ren, L. *Inorg. Chem. Commun*. 2004, 7, 572-575.

[136] Kuang, D.; Xu, A.; Fang, Y.; Liu, H.; Frommen, C.; Fenske, D. *Adv. Mater*. 2003, 15, 1747-1750.

[137] Jiang, Y.; Wu, Y.; Xie, B.; Yuan, S.; Liu, X.; Qian, Y. T. *J. Cryst. Growth* 2001, 231, 248-251.

[138] Wang, C.; Tang, K.; Yang, Q.; An, C.; Hai, B.; Shen, G.; Qian, Y. T. *Chem. Phys. Lett.* 2002, 351, 385-390.

[139] Li, C.; Yang, X.; Liu, Y.; Zhao, Z.; Qian, Y. T. *J. Cryst. Growth* 2003, 255, 342-347.

[140] Yang, J.; Zeng, J. H.; Yu, S. H.; Yang, L.; Zhang, Y. H.; Qian, Y. T. *Chem. Mater.* 2000, 12, 2924-929.

[141] Mo, M.; Zhu, Z.; Yang, X.; Liu, X.; Zhang, S.; Gao, J.; Qian, Y. T. *J. Cryst. Growth* 20003, 256, 377-382.

[142] Hu, H.; Mo, M.; Yang, B.; Zhang, X.; Li, Q.; Yu, W.; Qian, Y. *J. Cryst. Growth* 2003, 258, 106-112.

[143] Wang, D.; Yu. D.; Shao, M.; Yu, W.; Qian, Y. T. *Chem. Lett.* 2002, 1056-1057.

[144] Xie, Q.; Liu, Z.; Shao, M.; Kong, L.; Yu, W.; Qian, Y. T. *J. Cryst. Growth* 2003, 252, 570-574.

[145] Zhang, W.; Yang Z.; Huang, X.; Zhang, S.; Yu, W.; Qian, Y.; Jia, Y.; Zhou, G.; Chen, L. *Solid State Commun.* 2001, 119, 143-146.

[146] Shao, M. W.; Mo, M. S.; Cui, Y.; Chen, G.; Qian, Y. T. *J. Cryst. Growth* 2001, 233, 799-802.

[147] Zhang, H.; Yang, D.; Li, S.; Ma, X.; Que, D. *Nanotechnology* 2004, 15, 1122-1125.

[148] Zhang, H.; Ji, Y.; Ma, X.; Xu, J.; Yang, D. *Nanotechnology* 2003, 14, 974-977.

[149] Liu, Z.; Liang, J.; Li, S.; Peng, S.; Qian, Y. *Chem. Eur. J.* 2004, 10, 634-640.

[150] Kar, S.; Chaudhuri, S. *Chem. Phys. Lett.* 2004, 398, 22-26.

[151] Li, B.; Xie, Y.; Huang, J.; Su, H.; Qian, Y. *NanoStruc. Mater.* 1999, 11, 1067-1071.

[152] An, C.; Tang, K.; Liu, X.; Li, F.; Zhou, G.; Qian, Y. T. *J. Cryst. Growth* 2003, 252, 575-580.

[153] Li, W. J.; Shu, E. W.; Ko, J. M.; Chin, Z. Z.; Ogino, H.; Fukuda, T. *J. Crys. Growth* 2003, 250, 418-422.

[154] Lu, Q.; Gao, F.; Zhao, D. *Nano Lett.* 2002, 2, 725-728.

[155] Zhang, Y. C.; Hu, X. Y.; Qiao, T. *Solid State Commun.* 2004, 132, 779-782.

[156] Zhang, Y.; Li, Y. *J. Phys. Chem. B.* 2004, published on web A-G.

[157] Yang, J.; Xue, C.; Yu, S. H.; Zeng, J. H.; Qian, Y. T. *Angew. Chem. Int. Ed.* 2002, 41, 4697-4700.

[158] Zeng, J. H.; Yang, J.; Zhu, Y.; Liu, Y. F.; Qian. Y. T.; Zheng, H.G. *Chem. Commun.* 2001, 1332-333.

[159] Gautam, U. K.; Seshadri, R.; Rao, C. N. R. *Chem. Phys. Lett.* 2003, 375, 560-564.

[160] Chen, M.; Xie, Y.; Lu, J.; Xiong, Y.; Zhang, S.; Qian, Y.; Liu, X. *J. Mater. Chem.* 2002, 12, 748-753.

[161] Wu, J.; Jiang, Y.; Li, Q.; Liu, X.; Qian, Y. T. *J. Crys. Growth* 2002, 235, 421-424.

[162] Gao, F.; Lu, Q.; Zhao, D. *Chem. Lett.* 2002, 732-733.

[163] Bao, C.; Jin, M.; Lu, R.; Xue, P.; Zhang, Q.; Wang, D.; Zhao, Y. *J. Solid State Commun.* 2003, 175, 322-327.

[164] Zhang, H.; Ma, X.; Ji, Y.; Xu, J.; Yang, D. *Chem. Phys. Lett.* 2003, 377, 654-657.

[165] Tang, K. B.; Qian, Y. T.; Zeng, J. H.; Yang, X. G. *Adv. Mater.* 2003, 15, 448-450.

[166] Li, Y.; Liao, H.; Fan, Y.; Zhang, Y.; Qian, Y. T. *Inorg. Chem.* 1999, 38, 1382-1387.

[167] Gautam, U. K.; Rajamathi, M.; Meldrum, F.; Morgan, P.; Seshadri, R. *Chem. Commun.* 2001, 629-630.

[168] Xie, Y.; Wang,, W. Z.; Qian, Y. T.; Liu, X. M. *J. Solid State Chem.* 1999, 147, 82-84.

[169] Zhang, H.; Wang, L.; Xiong, H.; Hu, L.; Yang, B.; Li, W. *Adv. Mater.* 2003, 15, 1712-1715.
[170] Siemroth, J.; Henning, I. *Anal Chem Symp Ser*, 1981, 8, 339-348.
[171] Demarco, R. D.; Cattrall, R. W.; Liesegang, J. G.; Nyberg, L.; Hamilton, I. C. *Anal Chem.* 1992, 64, 594-598.
[172] Karanjai, M. K.; Dasgupta, D. *Thin Solid Films* 1987, 55, 309-315.
[173] Tohge, N.; Asuka, M.; Minami, T. *J. Non-Cryst. Solids* 1992, 147-149, 652-656.
[174] Yu, S. H.; Shu, L.; Yang, J.; Han, Z. N.; Quian, Y. T.; Zhang, Y. H. *J. Mater. Res.* 1999, 14, 4157-162.
[175] Murray, C. B.; Kagan, C. R.; Bawendi, M. G. *Science 1995, 270*, 1335-338.
[176] Braun, P.V.; Osenar, P.; Stupp, S. I. *Nature*, 1996, 380, 325-328.
[177] Birkmire, R.W.; Eser, E. *Annu. Rev. Mater. Sci.* 1997, 27, 625-653.
[178] Lotem, H.; Burshtein, Z. *opt. Lett.* 1987, 12, 561-563.
[179] Ozin, G. A.; Varaksa, N.; Coombs, N.; Davies, J. E.; Perovie, D. D.; Ziliox, M. *J. Mater. Chem.* 1997, 7, 1601-607.
[180] De Groot, K. *Bioceramics of calcium phosphate* Boca Raton: CRC Press, 1983.
[181] Lowenstam, H. A.; Weiner, S. *On biomineralization* Oxford, Oxford University Press, 1989.
[182] Tiselius, A.; Hjerten, S.; Levin, O. *Arch Biochem Biophys*, 1956, 65, 132-155.
[183] Bernardi, G. *Biochem Biophys Acta* 1969, 278, 423-434.
[184] Spencer, M.; Neave, E. J.; Webb, N. L. *J. Chromatogr.* 1978,166, 447-454.
[185] Buxbaum, G. *Industrial Inorganic Pigments* 1993, 1st ed.; VCH, Weinheim, Germany, 1993.
[186] Trovarelli, A. *Catal. Rev.*, 1996, 38, 439-520.
[187] Volger, J. *Physica* 1954, 20, 49-66.
[188] Hu, J. Q; Deng, B.; Zhang, W. X.; Tang, K. B.; Qian, Y. T. *Inter. J. Inorg. Mater.* 2001, 3, 639-642.
[189] Hu, J.; Lu, Q.; Deng, B.; Tang, K.; Qian, Y. T.; Li, Y.; Zhou, G.; Liu, X. *Inorg. Chem. Commu.* 1999, 2, 569-571.
[190] Xiao, J.; Xie, Y.; Tang, R.; Qian, Y. T. *J. Solid State Chem.* 2001, 161, 179-183.
[191] Hu, J. Q.; Deng, B.; Zhang, W. X.; Tang, K. B.; Qian, Y. T. *Inorg. Chem.* 2001, 40, 3130-3133.
[192] Liao, H. W.; Wang, Y. F.; Liu, X. M.; Li, Y. D.; Qian, Y. T. *Chem. Mater.* 2000, 12, 2819-2821.
[193] Lu, S. W.; Lee, B. I.; Wang, Z. L.; Samuels, W. D. *J. Cryst. Growth* 2000, 219, 269-276.
[194] Wang, C. R.; Tang, K. B.; Yang, Q.; Shen, G. Z.; Hai, B.; An, C. H.; Zuo, J.; Qian, Y. T. *J. Solid State Chem.* 2001, 160, 50-53.
[195] Wang, C. R.; Tang, K. B.; Yang, Q.; Hai, B.; Shen, G. Z.; An, C. H.; Yu, W. C.; Qian, Y. T. *Inorg. Chem. Commun.* 2001, 4, 339-341.
[196] Yan, L.; Li, Y.; Deng, Z. X.; Zhuang, J.; Sun, X. *Inter. J. inorg. Mater.* 2001, 3, 633-637
[197] Chen, Z.; Shi, E.; Li, W.; Zheng, Y.; Zhong, W. *Mater. Lett.* 2002, 55, 281-284.
[198] Cabanas, A.; Darr, J.; Lester, E.; Poliakoff, M. *J. Mater. Chem.* 2001, 11, 561-568.
[199] Zhu, D.; Zhu, H.; Zheng, Y. *Appl. Phys. Lett.* 2002, 80, 1634-1636.
[200] Zhu, D.; Zhu, H.; Zhang, Y. H. *J. Phys. Condens. Matter* 2002, 14, L519-L524.

[201] Heemskerk, J. W. T.; Noat, Y.; Bakker, D. J.; Ruitenbeek, J. M. V.; Thijsse, B. J.; Klaver, P. *Phys Rev B* 2003, 67, 115416.
[202] Zeng, H. C.; Lin, J.; Teo, W. K.; Wu, J. C.; Tan, K. L. *J. Mater. Res.* 1995, 10, 545-552.
[203] Lee, Y. P.; Kudryavtsev, Y. V.; Nemoshkalenko, V. V.; Gontarz, R.; Rhee, J. Y. *Phys. Rev. B* 2003, 67, 104424.
[204] Nin, H. I.; Chen, Q. W.; Lin, Y. S.; Zhu, H. F.; Ning, M. *Nanotechnology* 2004, 15, 1054-1058.
[205] Shen, S.; Hidajat, K.; Yu, L. U.; Kawi, S. *Adv. Mater.* 2004, 16, 541-545.
[206] Jaing, L.; Gao, L. *Chem. Mater.* 2003, 15, 2848-853.
[207] Yu, S. H.; Cui, X.; Li, L.; Li, K.; Yu, B.; Antonetti, M.; Colfen, H. *Adv. Mater.* 2004, 16, 1636-1640.
[208] Kang, M. *J. Mole. Catalysis A: Chem.* 2003, 197, 173-183.
[209] Kang, M.; Choung, S. J.; Park, J. *Catalysis Today* 2003, 87, 87-97.
[210] Yu, L.; Zhang, X. *Mater. Chem. Phys.* 2004, 87, 168-172.
[211] Hirano, M.; Ota, K.; Iwata, H. *Chem. Mater.* 2004, 16, 3735-3732.
[212] Yu, J. C.; Wu, L.; Li, P.; Li, Q. *Chem. Commun.* 2003, 1552-1553.
[213] Sorescu, M.; Tarabasanu-Mihaila, D. *J. Phys. Chem. Solids* 2004, 65, 1719-1725.
[214] Li, J.; Xu, Y.; Wu, D.; Sun, Y. *Solid State Commun.* 2004, 130, 619-622.
[215] Liu, C. M.; Zhang, D. Q.; Zhu, D. B. *Crystal Growth and Design* 2003, 3, 363-368.
[216] Vel L., Demazeau G., Etourneau J. *Mater. Sci. Eng. B* 1991, 10, 149-164.
[217] Jiang, L.; Fitzgerald, A. G.; Rose, M. J.; Lousa, A. *Appl. Surf. Sci.* 2000, 167, 89-93.
[218] Wentorf, R. H., DeVries R. C., Bundy F. P. *Science* 1980, 208, 873-880.
[219] Mirkarimi, P. B.; McCarty, K. F.; Medlin, D. L. *Mater. Sci. Eng. R* 1997, 21, 47-100.
[220] Talwar, D. N.; Sofranko, D.; Mooney, C.; Tallo, S. *Mater. Sci. Eng. B* 2002, 90, 269-277.
[221] Will, G.; Perkins, P. G. *Mater. Lett.* 1999, 40, 1-4.
[222] Park, K. T.; Terakura, K.; Hamada, N. *J. Phys. C: Solid State Phys.* 1987, 20, 1241-251.
[223] Will, D. R.; Keown, R. *J. Chem. Phys.* 1967, 47, 3113-119.
[224] Fitz, C.; Kolitsch, A.; Fukarek, W. *Thin Solid Films* 2001, 389, 73-179.
[225] Koga, H.; Nakamura, Y.; Watanabe, S.; Yoshida, T. *Surf. Coat. Tech.* 2001, 142, 911-915.
[226] Mishima, O.; Era, K.; Tanaka, J.; Yamaoka, S. *Appl. Phys. Lett.* 1988, 33, 962-964.
[227] Bhattacharya, A.; Botez, D. *Appl. Phys. Lett.* 1998, 72, 138-140.
[228] Well, R.L.; Aubuchon, S. R.; Kher, S. S.; Lube, M. S.; White, P. S. *Chem. Mater.* 1995, 7, 793-800.
[229] Trecce, R. E.; Macala, G. S.; Rao, L.; Franke, D.; Eckert, H.; Kaner, R. B.; *Inorg. Chem.* 1993, 32, 2745-752.
[230] Ashley, T.; Dean, A. B.; Elliott, C. T.; Pryce, G. J.; Johnson, A. D.; Willis, H. *Appl. Phys. Lett.* 1995, 66, 481-483.
[231] Carpenter, M. K.; Verbrugge, M. W. *J. Mater. Res.* 1994, 9, 2584-591.
[232] Godines, J. A.; Castillo, R.; Martinez, J.; Navarro, M. E.; De Anda, F.; Canalas, A.; Guzman, J.; Rios-Jara, D. *J. Cryst. Growth* 1997, 178, 422-425.
[233] Bieger, T.; *Microsystem Technol.* 1997, 3, 155-163.

[234] van Schnering H. G., honie W. *Encyclopedia of Inorganic Chemisty* Wiley, Chichester, UK, 1994, 6, 3106-206.

[235] Singh, R.; Sinha, S.; Chou, P.; Hsu, N. J. Radpour, F. *J. Appl. Phys.* 1989, 66, 6179-181.

[236] Wojtowicz, A.J. *Nucl. Instrum. Methods A* 2002, 486, 201-207.

[237] Cohen, M. L. *Phys. Rev. B* 1985, 32, 7988-991.

[238] Liu, A. Y.; Cohen, M. L. *Science* 1989, 245, 841-842.

[239] Cohen, M. L. *Science* 1993, 261, 307-308.

[240] Kornmann M., Funk R., *Aluminum* 1977, 53, 249-252.

[241] Pierson H. O. *Mater. Manuf. Process* 1993, 8, 519-534.

[242] Jones, M.; Horlock, A. J.; Shipway, P. H.; McCartney D.G.; Wood, J.V. *Wear 2001,* 250, 1009-016.

[243] Yu, M. Y.; Dong, S.; Li, K.; Hao, X.; Lai, Z.; Wang, Q.; Cui, D.; Jiang, M. *J Crys. Growth* 2004, 270, 85-91.

[244] Hao, X.; Dong, S.; Fang, W.; Zhan, J.; Li, L.; Xu, X.; Jiang, M. *Inorg. Chem. Communi.* 2004, 7, 592-594.

[245] Sardar, K.; Rao, C. N. R. *Adv. Mater.* 2004, 76, 425-429.

[246] Grocholl, L.; Wang, J.; Gillan, E. G. *Chem. Mater.* 2001, 13, 4290-4296.

[247] Gao, S.; Lu, J.; Zhao, Y.; Chen, N.; Xie, Y. *Eur. J. Inorg. Chem.* 2003, 1822-1827.

[248] Wei, S.; Yu, W. C.; Qian, Y. T. *J Appl. Phys.* 2004, 95, 3683-3688.

[249] Wei, S.; Lu, J.; Yu, W. C.; Zhang, H.; Qian, Y. T. *Chem. Lett.* 2004, 33, 396-387.

[250] Lu, J.; Wei, S.; Yu, W. C.; Zhang, H.; Qian, Y. T. *Inorg. Chem.*, 2004, 43, 4543-4545.

[251] Li, Y. D.; Duan, X. F.; Qian, Y. T.; Yang, L.; Ji, M. R.; Li, C. W. *J. Am. Chem. Soc.* 1997, 119, 7869-7870.

[252] Qian, Y. T. *Adv. Mater.*1999, 11, 1101-102.

[253] Li, Y.; Wang, Z.; Duan, X.; Zhang, G.; Wang, C. *Adv. Mater.* 2001, 13, 145-148.

[254] Xie, Y.; Lu, J.; Yan, P.; Jiang, X.; Qian, Y. T. *J. Solid State Chem.* 2000, 155, 42-45.

[255] Liu, J.; Chen, X.; Shao, M.; An, C.; Yu, W. C.; Qian, Y. T. *J. Cryst. Growth*, 2003, 252, 297-301.

[256] Cao, M.; Hu, C.; Wang, E. *J. Am. Chem. Soc.* 2003, 125, 11196-11197.

[257] Zhang, Z.; Biom, D. A.; Gai, Z.; Thompson, J. R.; Shen, J.; Da, S. *J. Am. Chem. Soc.* 2003, 125, 7528-7529.

[258] Ban, Y. J.; Lu, B.; Liu, Z, G.; Li, L.; Cui, D. L.; Xu, X. G.; Wang, Q. L. *J. Crys. Growth* 2003, 247, 503-508.

[259] Gu, Y.; Qian, Y.; Chen, L.; Zhou, F. *J. Alloys Comp.* 2003, 352, 325-327.

In: New Topics in Crystal Growth Research
Editor: George V. Karas, pp. 125-152
ISBN 1-59454-837-4

Chapter 3

GROWTH OF ALLOY NANOWHISKERS

Sin-iti Kitazawa
Dept. of Materials Science
Japan Atomic Energy Research Institute, Ibaraki, Japan.

ABSTRACT

In recent years, whiskers have become the subject of intense study since their remarkable characteristics, such as ultrahigh strength, low-density, high-modulus, high-melting point. Whiskers are metallic and non-metallic filamentary single crystals that have micrometer-sized widths or diameters. Furthermore, a large interest on "nanowhisker" that have nanometer-sized widths or diameters have been attracted with the advance on the nanotechnology. Growth of nanowhisker has become a necessary and important phase of nanotechnology since the nanowhisker changed from a laboratory curiosity to a most desirable structural component. For the growth of whiskers, it has been proven by several technique which include stress-induced growth, evaporation-condensation, vapor-phase reaction, growth from the melt, electrolytic growth, growth from solution and directional solidification of binary eutectics. Those scale down methods have been applied for nanowhisker growths. For the growth of alloy nanowhiskers, it has not been reported in great numbers, since it is more complicate than that of a simple substance. In this paper, a novel growth of the alloy nanowhisker is described. The carbon - nickel - indium mixture thin films were prepared by pulsed laser deposition. The nanowhiskers were synthesized by the annealing of the films. The component of the alloy was analyzed to be Ni_2In_3, whose melting point is slightly lower than the annealing temperature. The unique growth mechanism of nanowhiskers is explained the supersaturated liquid Ni-In alloys were seeping out from a porous graphite thin film.

1. INTRODUCTION

Whisker is, originally the meaning of the long stiff hairs that grow near the mouth of a cat, a metallic or a non-metallic single crystal fiber. Bulk is three-dimensional and thin film is two-dimensional, while whisker is a one-dimensional crystal. Whisker has been known as a natural and fibrous mineral, and it became a subject of study for an artificial crystal as a cause of short circuits of miniaturized communication equipment. Whisker was proven to appreciate that it increases the strength of materials, such as metal and plastic, by additive, since the density is small and the mechanical strength is very high. Thus whisker has become a subject of scientific study and engineering developments. For studying the history of "whisker", there are many textbooks for references. [Levitt 1970, Kaneko 1993, Fujiki and Mitomo 1993, Belitskus 1993]

With the advance of the nanotechnology, numerous attempts for material miniaturizations have been carried out. Various innovative phenomena have appeared by the size that not only miniaturization of the shape but also quantization, such as quantum size effects, quantum dot, etc. In optically, phenomena, originated from utilizing of tips whose sizes are smaller than the wavelength of light, was appeared. Furthermore, a bottom-up technique has taken the place of a top-down technique of large broaching, since a bottom-up technique has been noticed as a fabrication method of nanostructure materials using self-organizations that have possibilities to a high-volume production capability. [Timp 1999, Sakaki 2004]

In the study of nanostructured metal alloys, the approaches have been performed in the both top-down and bottom-up techniques. The milling method is a representation of the former, though it is not moderate for synthesis of any designed structures. For the latter, there are several methods, such as inert-gas evaporation of physical methods and coreduction from metal salt of chemical methods. In all cases, the prepared structures are powders or thin films. [Nalwa 2000, Nogi 2003]

In the study of nanowhiskers, carbon nanotube and semiconductor nanowire are representative. [Appell 2002, Ball 2001] Carbon nanotube is studied very well as a nano technology material, thus many methods have been applied to the generations in both science and engineering. [Iijima 1991] Several shapes of carbon nanotube are discovered, such as single-wall nanotube (SWNT) and multi-wall nanotube (MWNT). [Endo *et al.* 1996, Dresselhaus *et al.* 1996] Although, the possibility is limited, since they are only carbonic allotropes on balance. Semiconductor nanowire is also studied very well for the potential of applications in electronic devices. For example, new manufactures of layered semiconductor nanowires were reported using similar methods for the purpose of preparing one-dimensional diodes. [Gudiksen *et al.* 2002, Björk *et al.* 2002] These methods are using either laser ablation of solid target or chemical deposition methods, and the nanowhiskers were grown using the vapor-liquid-solid (VLS) method from a gold (Au) seed particle as about 10 nanometers in the diameter. The semiconductor vapor condensed around the seed and began to crystallize out between the seed and the surface in a small cylinder that raised the particle off the surface after all. A different precursor semiconductor was supplied in order to change the composition of the next bit of wire, and it crystallized between the seed and the previous semiconductor. By these methods, some of the most important types of semiconductors in the application following were produced, such as silicon (Si), silicon-germanium (SiGe), gallium arsenide (GaAs), gallium phosphide (GaP), indium arsenide (InAs), and indium phosphide (InP). [Service 2002] In these methods, by arranging tiny semiconductor tie in a row, though the prepared devices are not material manufactures with new forms.

The growths of nanowhiskers in the mechanisms except for VLS method have been also carried out. Cheng *et al.* (2002) produced nanowires of bismuth (Bi) with diameters ranging from 30 to 200 nm and lengths up to several millimeters using the residual stress of a chrome-nitride (CrN) thin film that deposited in magnetron sputtering. In this stress-induced method, the driving force is very larger than those of other methods, thus the growth rate is rapid as a few micrometers per second. Eventually the aspect ratio, the ratio length per diameter, is also large as at least 10^5.

In this manuscript, a novel method of growth of alloy nanowhiskers is described. The growth is performed by annealing of a mixture thin film produced by the laser ablation method, while the gas phase methods are often used for single materials like carbon nanotube, and VLS methods are used for semiconductors or other metal. In section 2, the experiments of growth of alloy nanowhiskers are described in detail. The experimental procedure contains the experimental apparatus and conditions. The experimental results contain emission spectra of laser ablation plasma, several micrographs of prepared specimens and various contents for the properties. In section 3, the growth mechanism of the alloy nanowhisker is introduced from the detail investigations of the experiments. Pulsed laser deposition (PLD) of thin films is mentioned, since the features of the mixture thin films prepared by PLD are important to understand the mechanism. In addition, the feature of the formed alloy is also inevitable . The growth model is proposed as considering the axial capillarity from the prepared thin film in the solution phase. In section 4, the possibilities of the growth of other alloy nanowhisker and proposals of the applications are described. The applications that might be enabled by the alloy nanowhiskers are taken up from the idea already advocated elsewhere. In section 5, the conclusion is mentioned.

2. EXPERIMENTS

2.1. Experimental Procedure

The experimental procedure is consisted of pulsed laser deposition (PLD), annealing and morphology. Fig. 1 shows a schematic drawing of the procedure. [Kitazawa *et al.* 2004a]

1) Laser

The laser used for PLD is a Nd:YAG laser (Infinity 40-100, Coherence). It was operated at the fundamental wavelength of 1064 nm, a pulse width of 3.5 ns, and repetition frequency of 20 Hz. The average energy per pulse was about 30 mJ, and the diameter of the laser spot was about 0.5 mm, thus an evaluated power density was 4 x 10^9 W/cm^2. [Kitazawa *et al.* 2003a] The targets are attached to a carousel type holder, since single laser beam can evaporate multiple materials alternatively. Fig. 2 shows a schematic drawing of multiple targets on the carousel type target holder. The holder can be rotated at the speed of 30 rotations per minute (rpm) for the purpose of alternative depositions. The materials of targets are graphite (C), nickel (Ni) and indium (In). The ratio of the area on the irradiated trace is C : Ni : In = 1 : 3 : 4. When the ratio changes to some degrees, no whiskers are produced. The dose of the evaporated particles is calculated by

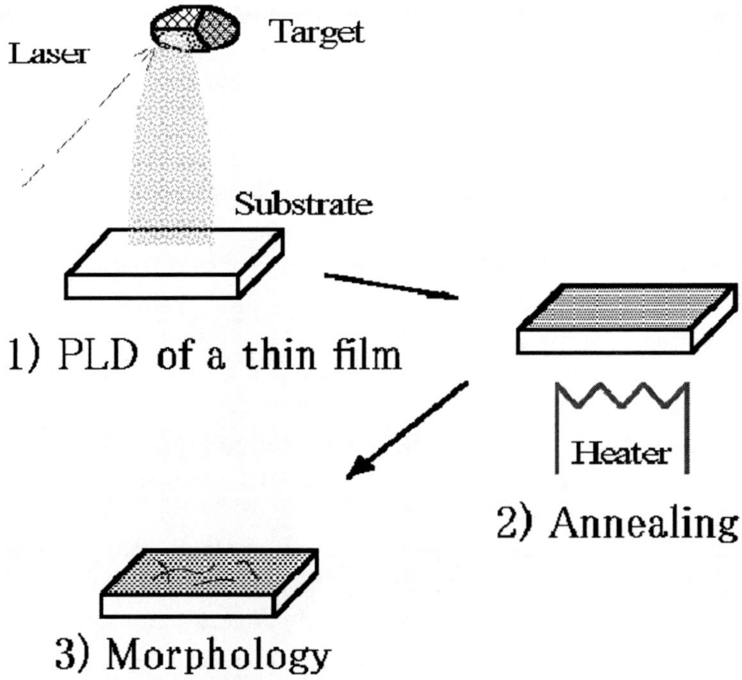

Figure 1. Experimental procedure of the growth of alloy nanowhiskers

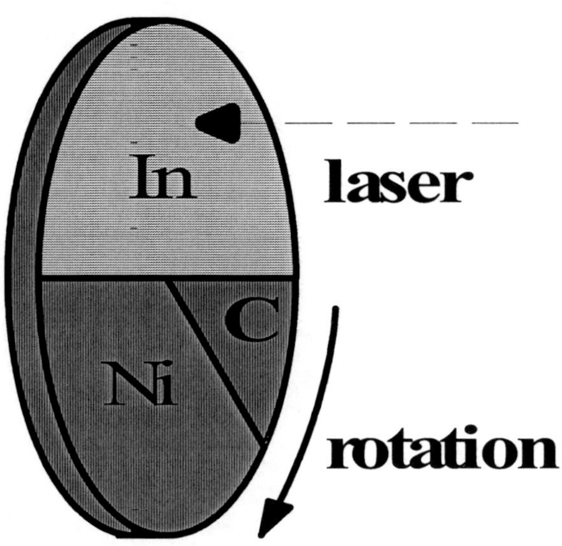

Figure 2. Schematic drawing of a carousel type target holder. As the targets (carbon, nickel and indium) rotate, they evaporate alternatively. The area of the targets controls the ratios of the evaporated quantities.

$$D = S \cdot Q \cdot N \qquad (1)$$

where Q is evaporated quantities per laser pulse, S is the area and N is the number of laser pulses. However, it should be considered analytically, since evaporated quantities easily vary with energy density, wavelength, pulse duration, etc. [Chen 1994, Singh and Narayan 1990]

Furthermore the evaporated quantity is very complicated, since the distribution of the deposition flux has cosθ dependence for normal line angle θ which differs in each material. [Saenger 1994]

2) Deposition Chamber

PLD of the thin films were performed in a vacuum chamber. Fig. 3 shows the PLD system that was used in this study. The base pressure of the chamber was about $4x10^{-4}$ Pa before and after the depositions. However, it cannot be measured during the deposition owing to the existence of ablation plasma in the space between the substrate and the target. The atmosphere gas seems to be fewer, since getter effects arose by the ablation plume. The distance between the target and the substrate was 2.5 cm in normal line direction. It is possible that the temperature of the substrate is heated from the room temperature to 400 °C by an attached heater and temperature sensor at the back of the substrate holder. In the chamber, the laser beam was introduced through a glass window at an angle of about 45 degrees for the target. It is possible to observe the laser plume through the glass window from vertical direction.

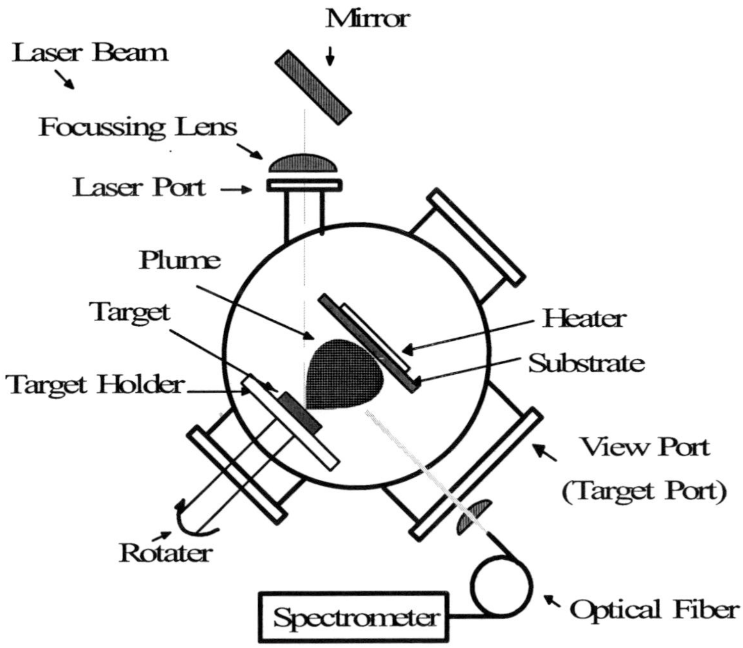

Figure 3. Experimental configuration of PLD system for thin film preparation with optical emission spectroscopy of the ablation plume.

3) Substrate

Several materials were used for the substrate of PLD. A successful one is MgO(100). The motivation of using MgO(100) is the high temperature stability and hardness as substrates. There is also a good relation with nickel and MgO(100), which is not so important in this study. [Lin *et al.* 2003, 2002] Prior to PLD, the MgO(100) plates were sufficiently washed the surface in acetone and ethanol. KBr substrates were prepared for transmission electron microscope (TEM) observations, since it is easy to tear off the deposited thin films in water. The temperature of the substrates during PLD was kept in the room temperature (~ 20 °C) or heated to an objective temperature. In the room temperature, it was heated by the energy of the deposited particles, and then the temperature rose to about 30 °C.

4) Deposition and Annealing

The typical thickness of a thin film deposited for 8 minutes with growth of alloy nanowhisker is almost 20 nm. The deposition speed is calculated as one laser pulse deposited 0.002 nm per pulse in average. The film was not layered but mixed, since the thickness of film deposited by one pulse is smaller than Bohr radius, 0.053 nm. The volume is affected by the laser power density, repetition frequency, and so on, though it is a good condition for some unclear reasons. Annealing was performed to the deposited thin film on the substrate in the deposition chamber at the base pressure after PLD or in another oven at the atmosphere. The experimental conditions above mentioned are summarized in Table 1.

Table 1. Experimental conditions

Laser	
Apparatus	Infinity 40-100, Coherence
Wavelength	1064 nm
Pulse width	3.5 ns
Repetition frequency	20 Hz
Averaged energy per pulse	30 mJ
Spot diameter	0.5 mm
Evaluated power density	4×10^9 W/cm^2
Target	
Materials	Carbon - Nickel - Indium
Holder	Carousel type
Rotation speed	30 rpm
Target - substrate distance	25 mm
Substrate	
Material	MgO(100), KBr, etc.
Temperature during deposition	RT(20 °C) ~ 400 °C
Annealing	
Temperature	~ 400 °C
Time	1 h
Environment	Base pressure or atmosphere
Others	
Base pressure	4×10^{-4} Pa
Deposition time	8 minutes
Film thickness	about 20 nm
Deposition speed	0.002 nm/pulse

5) Morphology

The surface morphology was performed by a scanning electron microscope (SEM) and a transmission electron microscope (TEM). In each microscope, energy dispersive method X-ray spectroscopes (EDS) are installed. The SEM operated was a general purpose SEM (JSM-5600, JEOL). [Kitazawa 2004, Kitazawa *et al.* 2004b] The resolution is about 5 nm, when the acceleration voltage was 15 kV. The TEM is an analytical electron microscope that offers micrographs and diffraction patterns (JEM-2000FX, JEOL). It was operated at the acceleration voltage of 200 kV. Furthermore, an atomic force microscope (AFM) (JSPM-4200, JEOL) was operated to observe the surfaces, though no significant image has observed except for some scratched trails.

6) Crystallography

In the crystallographic verification of the prepared thin films, micro-Raman spectroscopy and X-ray diffraction (XRD) method were employed. For micro-Raman spectroscopy, an Ar ion laser was operated with wavelength of 514 nm as incident beam, which provided a spot size of the order of 1 μm. The spectrum was observed in the range from 1000 to 1800 cm^{-1}. [Kitazawa *et al.* 2005] For XRD, the crystallographic relationships between the thin films and the substrates were investigated with a high-resolution X-ray diffractometer (X'Pert-MRD, Philips). The X-ray source was operated at 40 kV and 30 mA for the Cu-Kα line. The orientations of the hetero-epitaxial films were examined using θ-2θ scans and in-plane ϕ scans. [Kitazawa 2004, Kitazawa *et al.* 2004b, 2003b]

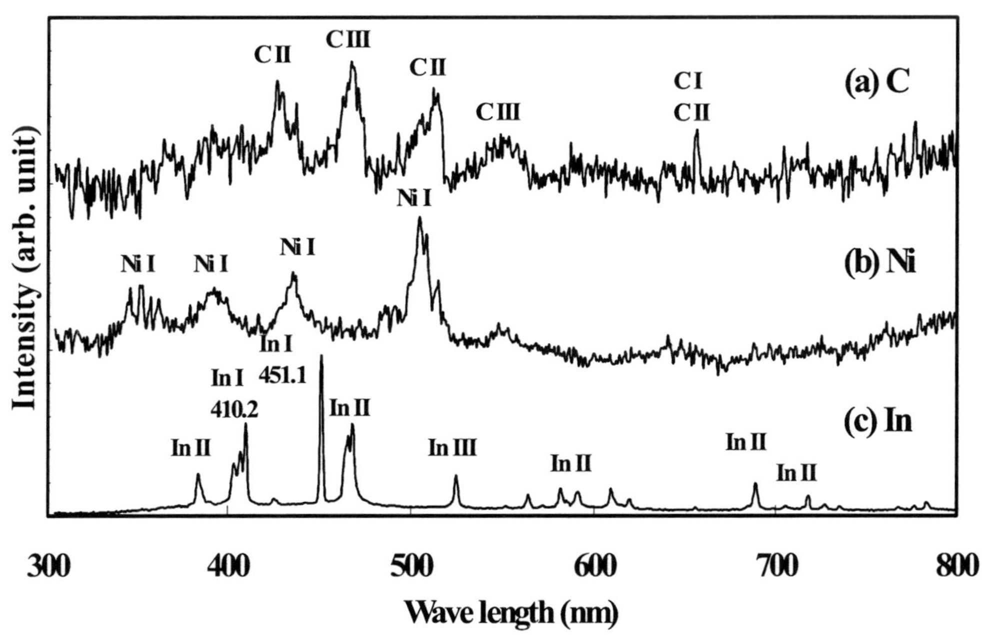

Figure 4. Optical emission spectra from the ablation plume in PLD of the mixture thin film. The target of upper spectrum (a) is graphite (C), the middle (b) is nickel (Ni) and the lower (c) is indium (In). The spectra were observed in the alternative ablation using the carousal type target holder.

2.2. Experimental Results

1) Emission Spectroscopy of Laser Ablation Plume

Fig. 4 shows the optical emission spectra of the laser ablation plumes. In the condition of these plumes, it is possible to estimate the plasma temperature by Saha equation with assuming the establishment of a local thermal equilibrium (LTE) condition. [Kitazawa 2004, Geohegan 1994, Yamamoto and Murayama 1995, Chen 1974]

From Saha equation, the ratio of the spectral intensity is calculated,

$$\frac{I_i}{I_k} = \frac{\nu_i A_i G_i}{\nu_k A_k G_k} \cdot \exp\left(\frac{-(E_i - E_k)}{kT}\right) \tag{2}$$

where I_i and I_k are the spectral intensities, ν_i and ν_k are the transition frequencies and A_i and A_k are the spontaneous emission coefficients. G_i and G_k are the degeneracy of the upper state, E_i and E_k are the energies of the transition, k is Boltzmann constant and T is the temperature of the plasma. Some peculiar lines of atomic carbon (C), nickel (Ni) and indium (In) are assigned and denoted with the wavelengths and the Roman numeral notation. [Martin *et al.* 1999, Bashkin and Storner 1978a, b, Kitazawa and Yamamoto 2003b] In the notation, the spectrum emitted by neutral atoms of a given element is called the first spectrum of that element, and is denoted by the Roman numeral I; the spectrum emitted by singly ionized atoms is called the second spectrum and is denoted by Roman numeral II; etc. Most of all peaks are not possible to describe into the fine structure, as the resolution of the spectrograph (< 2 nm) is larger than the separation (< 1 nm). However, in case of the indium, In I (410.2 nm) and In I (451.1 nm) are very sharp and well isolated from other peaks. Substituting the value of the two peaks for equation (2), the plasma temperature is calculated to be 2.6×10^4 K $\approx$ 2.2 eV. However, the exact time resolution cannot be carried out, since fluorescent period of the ablation plume is about 100 ns and the exposure time of the spectroscope is 19 ms. Therefore, the temperature corresponds to a mean value of observed plasma.

2) Scanning Electron Microscope (SEM) Observation

The SEM was operated to evaluate features of the surface of the perpetrated thin films. Figs. 5~ 7 show the SEM images. Fig. 5 shows the surface of the C-Ni-In mixture thin film deposited at the room temperature. Many droplets, which are potsherds of the target flied directly, are observed on the surface clearly. They are peculiar "debris" in laser ablation. The particles with the diameter of 10 μm are droplets of indium, and the grains of the diameter under 1 μm are those of nickel. By EDS analysis, the widely distributed droplets are analyzed to pure indium and nickel metals. By micro Raman spectroscopy, the rectangle grains over 10 μm are analyzed to minute graphite crystals. Just after the deposition, the slight mass transfer might be caused. However, alloying or any construction did not happen either in the condition of as deposited.

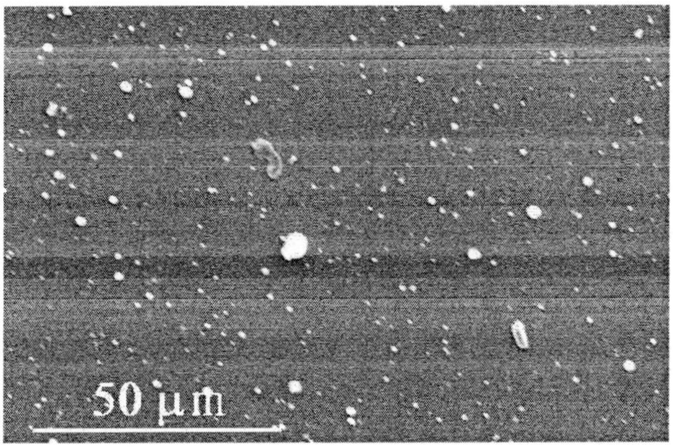

Figure 5. SEM micrograph of the surface of the carbon, nickel and indium (C-Ni-In) mixture thin film deposited on a MgO(100) substrate without any post-treatments. SEM was operated at the acceleration voltage of 15 keV. PLD was performed with the laser (Nd: YAG 1064 nm, 10 Hz, 4 x 10^9 W/cm^2) at room temperature. There are no whiskers on the surface.

Fig. 6 shows a SEM micrograph of the surface of prepared C-Ni-In mixture thin film after the annealing at 400 °C in 4 x 10^{-4} Pa for 1 hour. Several nanowhiskers are observed on the thin film. The lengths of the whiskers are less than 100 μm, and the diameters are from 5 to 50 nm. The typical aspect ratio was calculated with 2500 for a nanowhisker with diameter of 40 nm and length of 100 μm. There are some wires, whose sides have contrasts of double line, thus they may be structures of cavity. The whiskers were formed, only when the temperatures of the substrate during PLD were RT and 100 °C, but not 200, 300 and 400 °C. The composition ratio of the nanowhisker was analyzed by EDS, and the ratio of the atomic weight is Ni : In =2 : 3. This metal mixture becomes a Ni_2In_3 binary alloy that was stabilized less than 410 °C, as it is shown in the discussion section.

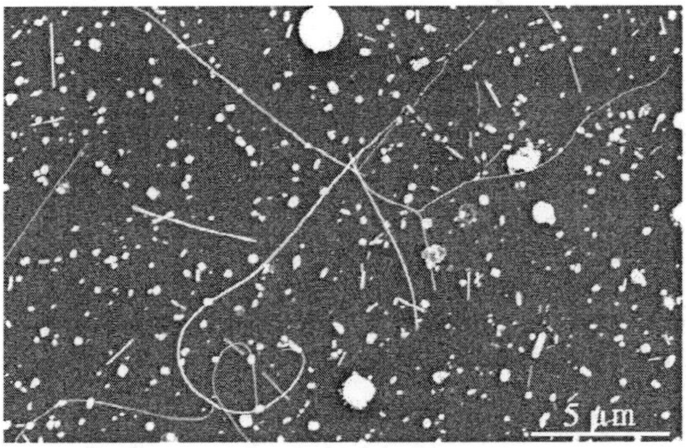

Figure 6. SEM micrograph of the surface of carbon, nickel and indium (C-Ni-In) mixture thin film deposited on a MgO(100) substrate with post-annealing. SEM was operated at the acceleration voltage of 15 kV. PLD was performed with the laser (Nd: YAG 1064 nm, 10 Hz, 4 x 10^9 W/cm^2) at room temperature. The annealing was done for 1 h at 400 °C in the base pressure of 4 x 10^{-4} Pa. There are a lot of whiskers on the surface.

Fig. 7 shows some other SEM micrographs with alloy nanowhiskers. In Fig. 7(a), there are some carbon debris with about 1 μm squares and some indium lump of round shapes. In Fig. 7(b), the part looks like a twisting paper is based on the whisker of multi-pieces. The blurred part of the whisker slightly vibrated actually by the electron irradiation of the SEM operation, and it was dissolved by long-time observation in the electron irradiation. In Fig. 7(c), it is the part in which the density of whiskers is high. Since there are some extent dispersions by the disposition of PLD, those of whiskers are also extent. Fig. 7(d) is the wide range image. Large things over 1 μm are debris. It is proven that an alloy nanowhisker is not possible to grow near indium debris, since the indium solution collects bordering small indium particle to enlarge the volume.

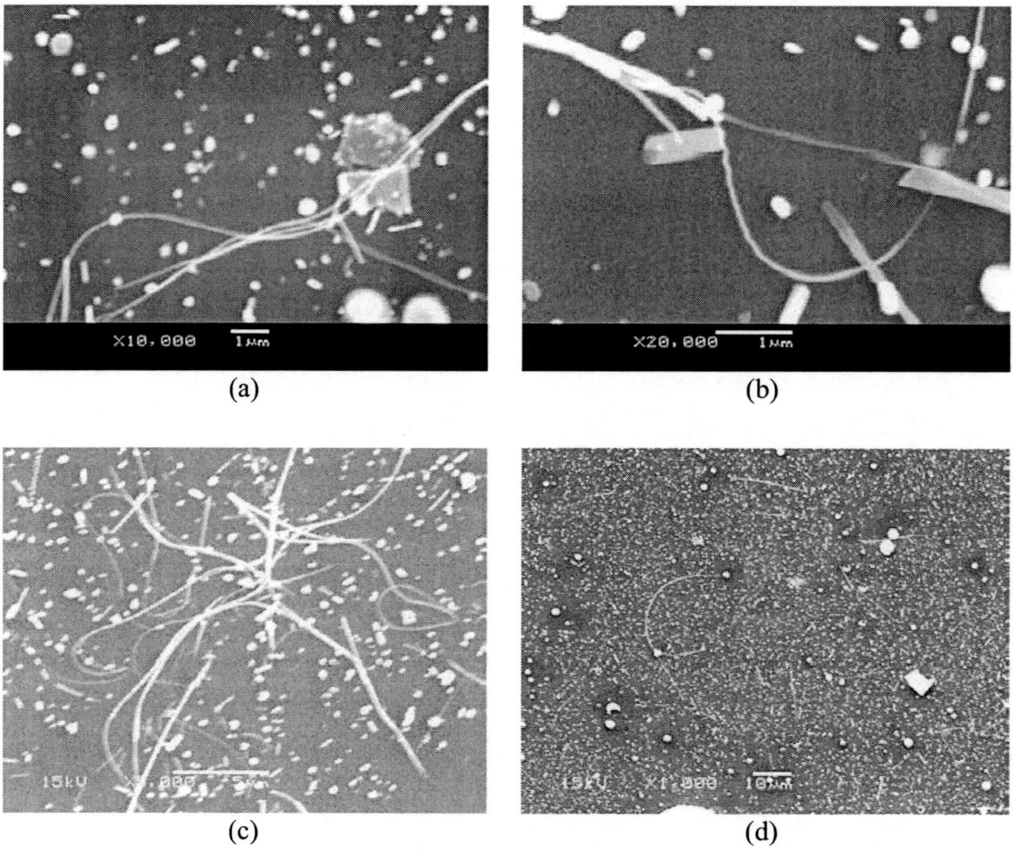

Figure 7. SEM images of carbon - nickel - indium mixture thin films with the alloy nanowhiskers. The growth parameters are the same as Fig. 6

3) Transmission Electron Microscopy (TEM) Observation

Fig. 8 shows a TEM micrograph of a C-Ni-In mixture thin film torn off from a KBr substrate. The whiskers are grown by annealing at 400 °C for 1 h. There is a nanowhisker of the diameter about 20 nm on the porous graphite thin film. The quantity of nanowhisker on a KBr substrate is overwhelmingly fewer than that on a MgO substrate. The KBr substrate was used to prepare the specimen for TEM observation. KBr substrate dissolved by water, thus the deposited thin film was easy to peal off by soaking in purified water. The film was supported on a cupper mesh grid. It applicative tried for hardened rock salt (NaCl) substrates that irradiated by gamma ray, though no whisker was observed by some breaches of some parameters.

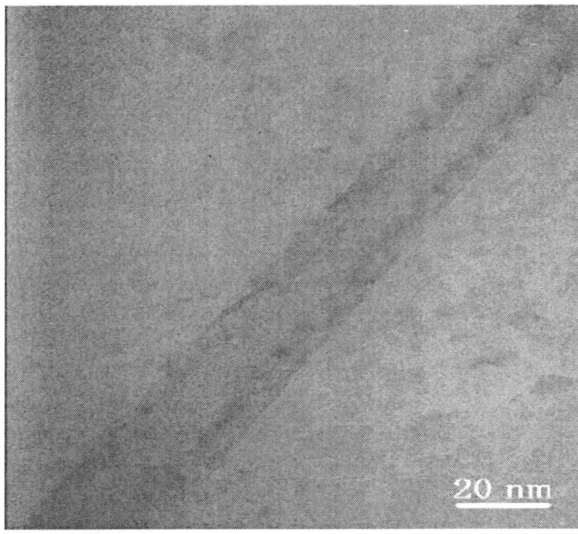

Figure 8. TEM micrograph of the prepared thin film with an alloy nanowhisker. The image from the right upper to the left downward is the alloy nanowhisker. TEM was operated at the acceleration voltage of 200 kV. The specimen was prepared by PLD and stripped off from a KBr substrate in purified water. PLD was performed with the laser (Nd: YAG 1064 nm, 10 Hz, 4 x 10^9 W/cm^2). The annealing was done for 1 h at 400 °C in the base pressure of 4 x 10^{-4} Pa

Fig. 9 shows the selected area diffraction image of the whisker in Fig. 8. The ring diffraction patters and arrays of spots were observed in the image. The brightest spot at the center was shielded to avoiding over exportation. The camera distance L is 80 cm and the acceleration voltage is 200 kV, thus the wavelength of electrons λ is calculated to be 0.0025 nm. The identification of crystallographic orientations was carried out from the observed ring diffraction image. Table 2 shows the comparison of the experimental results and references. R are radius of the rings on the photographic image, d_{obs} are the observed lattice spacing, d_{table} are calculated lattice spacing cited from references, and hkl are Miller indices. The reference data is sited from Pauling data. [Villars *et al.* 2001] The notations are followed the reference [Ashcroft and Mermin 1987] The diffraction pattern from graphite cannot be observed clearly, since the rings are broader and no speckle of an amorphous material was observed. [Williams and Carter 1996] The results from the ring diffractions are well agreed with Ni_2In_3, but not $Ni_3In_{6.4}$ or graphite. For that reason, the constitutional elements of the nanowhisker are proven to be Ni_2In_3 crystals. In addition, the spots of (hkl) = (101) and (202) were observed clearly from the point arrangement. Some angles of the spot arrays were not observed at 60 degrees, which is the expect angle for a hexagonal crystal, Ni_2In_3, as shown in the discussion section. It is explained that the electron beam irradiates two single crystals of the nanowhisker in an equal direction. That is to say, these two crystals are both sides of hollow structured tube. Fig. 10 shows the imagined drawing of the TEM observation of the alloy nanowhisker.

Figure 9. TEM observed diffraction image of the nanowhisker in Fig. 8.

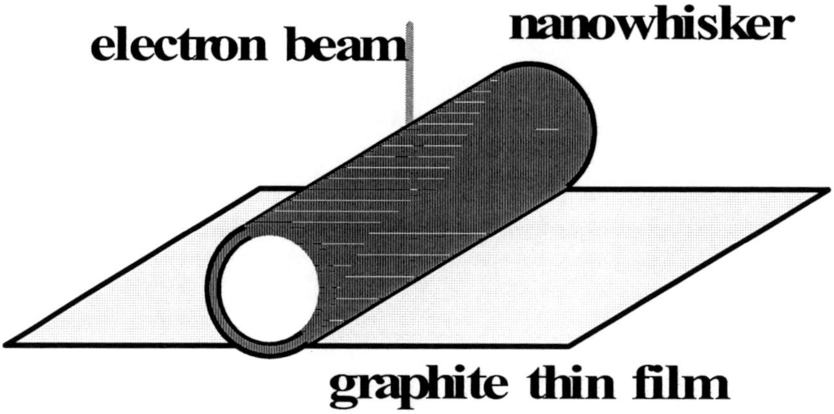

Figure 10. Schematic drawing of electron beam irradiation to the nanowhisker on graphite thin film. If the structure of nanowhisker is a hollow pipe state, there are two separate crystal planes on a thin film.

Table 2. Identification of Ni_2In_3 from the ring diffraction patterns

R (mm)	Intensity	d_{obs} (nm)	d_{table} (nm)	hkl
6.7	strong	0.30	0.30869	101
8.9	very strong	0.23	0.21935	110
9.6	weak	0.21	0.21721	012
13.9	weak	0.14	0.15434	202
16.1	medium	0.12	0.12623	122
17.1	weak	0.11	0.10968	220

4) Raman Spectroscopy of Thin Films

Fig. 11 shows a micro-Raman spectrum of the prepared mixture thin film after the annealing. There are two carbonic peaks in the range of 1000 - 1800 cm^{-1} in the spectrum. The centers of the peaks are estimated by a deconvolution using the Lorentzian, accordingly that of D-band is 1372 cm^{-1} and that of G-band is 1574 cm^{-1}. The intensity ratio is calculated; $I_{(D)}/I_{(G)} = 0.93$. It shows that the material of the thin film is mainly graphite and amorphous carbon on amorphous state. [Ferrari and Robertson 2000] However, metals are hard to detect in the Raman spectroscopy. The spectrum is hardly changed by the annealing, since a carbon thin film is practically stable for annealing at 400 °C. However nickel (Ni) and indium (In) were deposited with graphite (C) uniformly, it is also considered that the metals were doped in the graphite thin film.

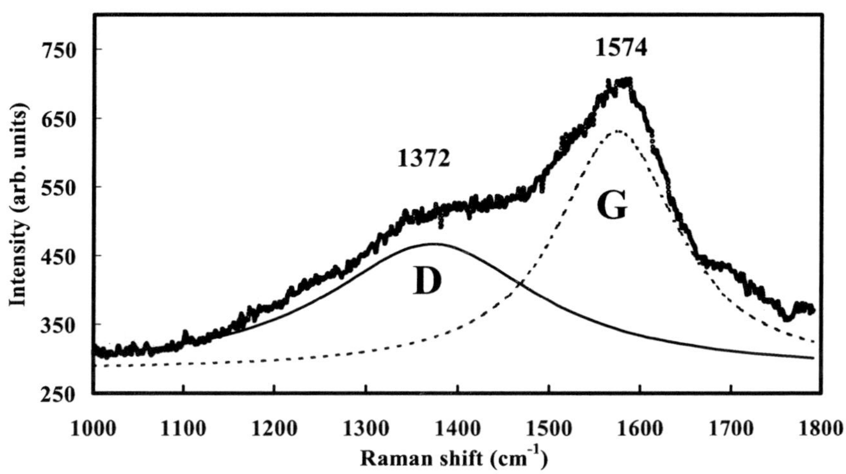

Figure 11. Raman spectrum of the carbon (C), nickel (Ni) and indium (In) mixture thin film. The dotted line presents measured value. The spectrum is deconvoluted by Lorenz curves. The solid line is the peak of D-band and the dashed line is that of G-band.

5) Growth Condition of the Nanowhisker

It is interesting to consider the experimental results in which the whiskers were not grown in order to examine the generation mechanism of the alloy nanowhisker. The conditions that no whisker growth was observed are showed as fellow.

i) C-Ni-In in which one is not sufficient

In the case that only two of the materials, such as C-Ni, C-In and Ni-In, were deposited, no whisker was grown. The experiments were not performed in a variety of parameters, since any motivating results were not expected.

ii) The film thickness were different

The nanowhiskers were not grown in which the thickness of the prepared thin film is reduced to the half (10 nm) or increased to five times (100 nm). However in the later case, some interesting patters of alloy particles were produced on the thin film. Fig. 12 shows some of the SEM micrographs. The film thickness were controlled by the duration time of PLD, namely the film of 10 nm in thickness were prepared for 4 minutes and that of 100 nm were for 40 minutes. The area except for the debris is C - Ni - In mixture. In Fig. 12(a), the nanometer-sized particles appeared and the micrometer-sized indium droplets disappeared on the surface, since the contaminations of indium metal were melted in the post-annealing. Fig. 12(b) shows that a pattern with a round particle consist as Ni: In = 4: 6 ~ 1: 2, diameter of 2.5 μm in the center. It is interesting that small particles are not present in the circumference. The annealing produced the center particle, since no alloy exists in the debris. Fig. 12(c) shows that various nanometer-sized particles consist a round shape of 9 μm in diameter, as surrounded by the wall of particles of Ni: In = 1: 3 circumference. The composition ratio of the center particle with the diameter of 1 μm is Ni: In = 1: 2.4. Fig. 12(d) shows that a micrometer-sized particle (0.5 μm, Ni: In = 2: 3) in the center is surrounded by nanometer-sized particles. The sizes of nanometer-sized particles are less than 100 nm in the inner and around 100 nm in the outer circumference. From the above surface morphologies, it is proven that annealing formed the patterns and the localization of contents ratio of the particles was produced in the surface. The differences in the forming of particles are depended on the heterogeneity of the deposited contaminations.

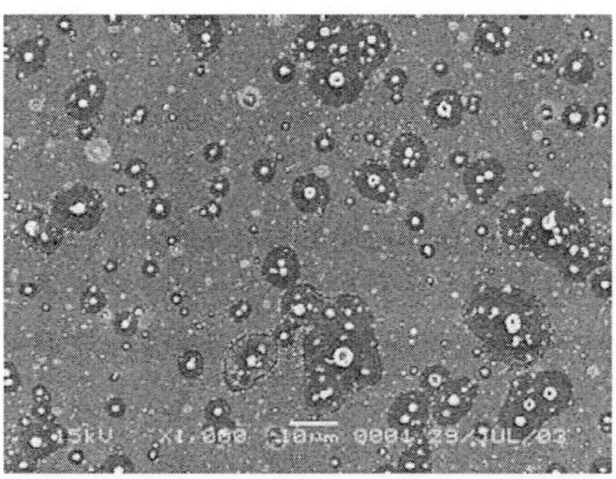
(a)

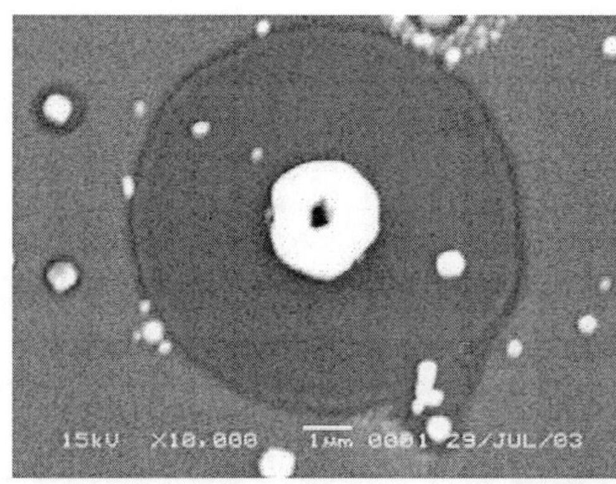

(b)

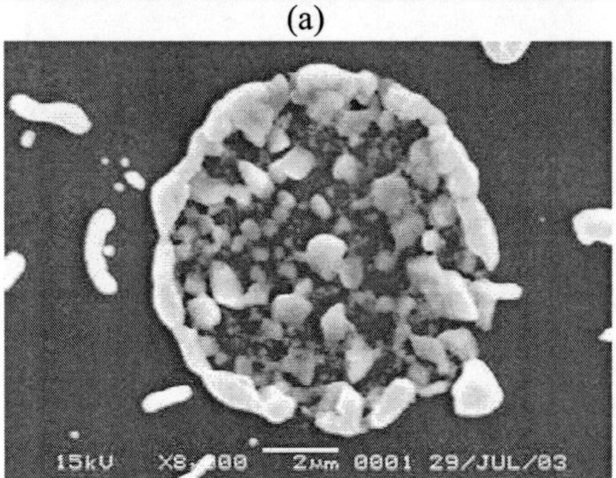

(c)

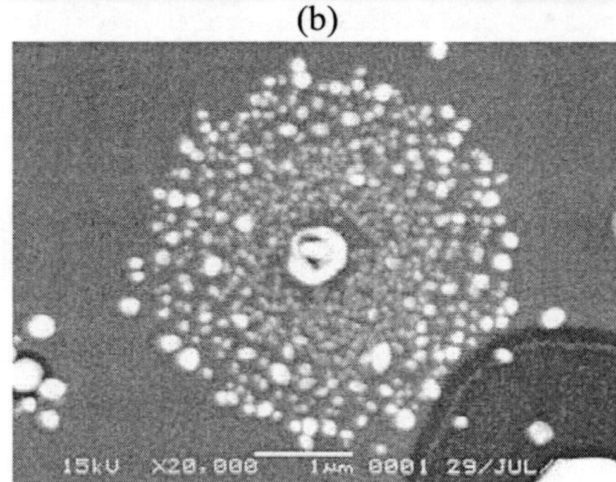

(d)

Figure 12. SEM images of interesting patterned alloys
(a) Low magnification SEM micrograph of the surface of carbon, nickel and indium mixture thin film after annealing. The scale bar indicates for 10 μm. (b) A pattern with a round particle of Ni-In alloy. The scale bar indicates for 1 μm. (c) A circular pattern consists of nanometer-sized Ni-In alloy particles. The scale bar indicates for 2 μm. (d) A firework like pattern consists of nanometer-sized Ni-In alloy particles. The scale bar indicates for 1 μm.

iii) Glass substrate

No whisker was observed when glass was used for substrate. It is assumed that a glass substrate eased the internal stress of thin film by the annealing at 400 °C. The reason is not cleared yet, though some reasons are taking into account, such as expansion coefficients, elasticity and surface tensions.

iv) Substrate temperature during PLD

The growths of nanowhiskers were observed in the case that the substrate temperature was kept at RT and 100 °C during PLD, but not at 200, 300 and 400 °C. The difference is based on the melting point of indium 156.6 °C. When indium melts during PLD, it forms a droplet and the inner stress originated from PLD is disappeared. It follows that the indium particles do not move by the post annealing and there is no alloying. It is very important for nickel and indium sufficient to disperse on the mixture thin film. It will be explained by the whisker growth mechanism as shown in the next section.

3. DISCUSSION

3. 1. Pulsed Laser Deposition of Thin Films

In the growth of alloy nanowhiskers, the inner stress of the prepared thin film is important. Therefore the mechanism of pulsed laser deposition (PLD) is presented as furnishing with much information. The more detail discussions are described in the references. [Chrisey and Hubler 1994, Singh and Narayan 1990, Hotta *et al.* 1999]

3. 1. 1. Mechanisms of PLD

PLD of thin films is one of the physical vapor deposition methods (PVD), which use the laser as a vapor source of the target material. In this method, an ablation locally is caused by irradiating of a high-power pulse laser onto a target material surface and instantaneously evaporates atoms, molecules, ions and clusters of the constituent element, which are pile up on a substrate. This method is also called as laser sputtering method, laser ablation deposition, pulsed laser ablation (PLA), etc.

The fundamental process of PLD of thin films is divided into three processes, such as laser ablation on a target, flight of the sputtered particles and film growth on a substrate. Fig. 13 shows the process.

Laser ablation is apparently a simple phenomenon that a high energy laser irradiates on materials and evaporates them, nevertheless, in the elementary process, there are still many unsolved problems, since some complicated mechanisms are contained. One of the evidences is the application in Koichi Tanaka's Nobel prized work, "a soft adsorption ionization method for mass spectrometric analysis of biological macromolecules" in his Nobel Lecture "The Origin of Macromolecule Ionization by Laser Irradiation". [Tanaka 2002]. He discovered "Soft laser desorption" method, in which proteins can be ionized by laser ablation adding a chemical/physical matrix to the target.

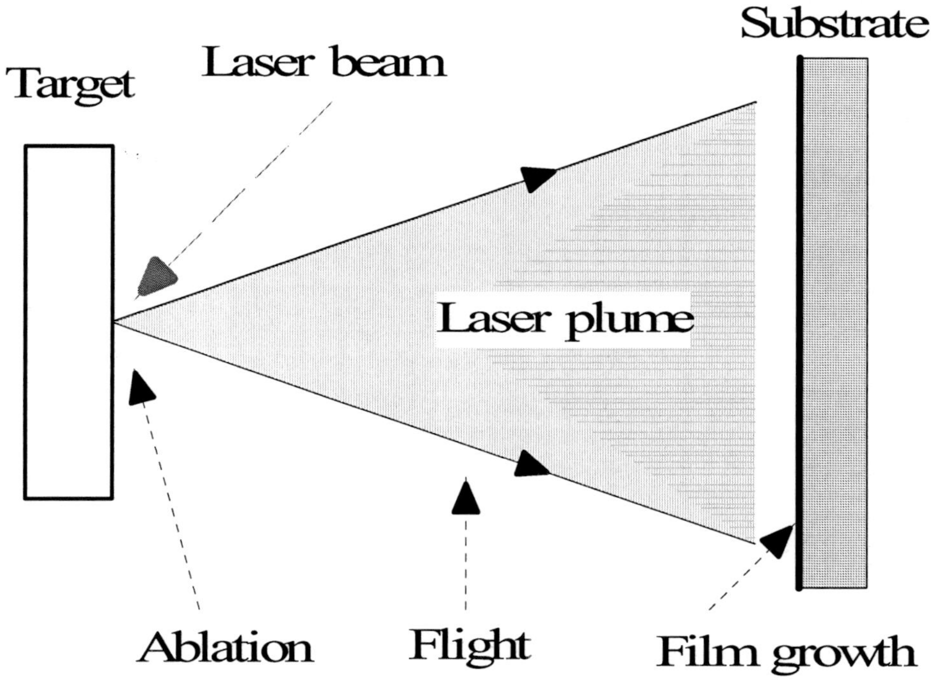

Figure 13. Fundamental process of PLD of a thin film
1) laser ablation, 2) flight of the sputtered particles, 3) film growth

The primary mechanism of the ablation is divided into absorption of photons, electronic excitation, emission of particles. It is possible to consider the evaporated particles in kinematics as ablation plasma. Fig. 14 shows the schematic drawing of the primary mechanism of laser ablation.

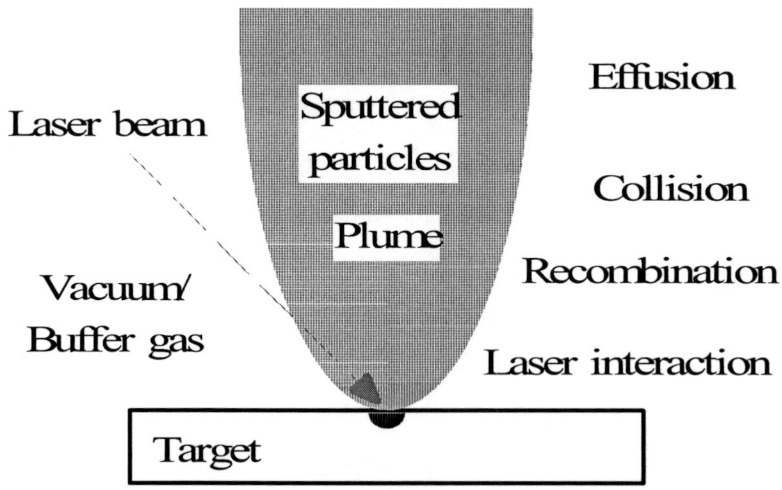

Figure 14. Mechanism of laser ablation

In the process of "absorption of photons", there are two different processes according the quantities between the laser frequency ω_{laser} [m] (= $2\pi \cdot$ c (speed of light) / λ_{laser} (laser wavelength)) and the metallic plasma frequency ω_p [m]. For the laser whose laser frequency ω_{laser} is higher than the plasma frequency ω_p, the reflectivity greatly decreases, and it would be able to penetrate into inside. While the laser, whose laser frequency ω_{laser} is below the plasma frequency ω_p, is all reflected on the metal surface and cannot penetrate inside, since the free electrons force to vibrate following the electric vector of incident light and cancel the electric vector vibration in solid. [Kelly and Miotello 1994, Murakami 1999] Generally metallic plasma frequency ω_p is given by

$$\omega_p = \left(\frac{4\pi ne^2}{m} \right)^{1/2}. \tag{3}$$

where n is the density of electrons, e is the unit electric charge and m is the mass of electron. [Chen 1974, Ashcroft and Marmin 1976] It is correspondent to the ultraviolet (UV) light for metals. In this study, the laser frequency is $\omega_{laser}/2\pi = 0.28 \times 10^{15}$ Hz for Nd:YAG 1064 nm, and the plasma frequencies of nickel and indium are $\omega_{nickel}/2\pi = 3.8 \times 10^{15}$ Hz and $\omega_{indium}/2\pi = 3.1 \times 10^{15}$ Hz, respectively. Therefore the laser cannot penetrate into inside of the targets. Nevertheless for the laser of high energy density there is another mechanism, which is called a thermal sputtering. While the laser energy density E_{laser} [J/cm^2] is higher than the threshold E_o [J/cm^2] that is determined by the target materials and the laser, a laser ablation is produced. In this case it is a dynamic process, which is followed by electron excitation and thermal heat-up. The irradiated photons also accelerate the electrons by inversed bremsstrahlung. Those processes take only a few picoseconds, thus the energy of photons is sequentially transferred to the thermal energy of the target during the laser irradiation of nanoseconds. [Kelly and Miotello 1994]

The secondary mechanisms of the ablation are consisted of plasma heating and free flight. The plasma absorbs the energy of the laser by the inversed bremsstrahlung during the interaction with the laser pulse. It takes only a few picoseconds that electrons, atoms and ions are emitted from solid surfaces after laser irradiation, thus the high-density plasma interacts with laser ray near the ablation point in the present study of nanosecond laser. The ionization of neutral atoms and ions are repeatedly promoted elasticity and inelastic collision by the collision with the accelerated electron by the inversed bremsstrahlung. This plasma state is called "plume", which has perpendicularly the distribution $\cos^n\theta$ from the ablation point on the surface. In the distribution $\cos^n\theta$, there is the dependence by mass of the target, laser intensity, etc. [Saenger 1994, Okada 1999] The light of longer wavelength is easier to interact by the inversed bremsstrahlung of plasma and laser beam. [Hughes and Anderson 1989, Kitazawa *et al.* 1993] The plasma temperature can be estimated by equation (2), though the translation velocity, which is accelerated in the secondary mechanisms, is not included.

The flight process starts in the secondary mechanism of the ablation. The sputtered particles have free flight in vacuum, since the plasma density is too small to interact. The excited species, such as atoms, molecules, ions and clusters accelerated in the ablation plume, fly at high-speed (~ 10^6 cm/s) with high kinetic energy (several keV) in vapor phase. If buffer gas is used in the atmosphere, it is not free flight and nanostructure may be prepared. [Kitazawa *et al.* 2004b, Kitazawa 2004]

In the film growth process, the excited species pile up on the vapor deposition substrate, and a thin film grows with crystallization. It is known that there is large residual stress in a thin film according to the peening effect, since depositing particles have high kinematics energies. [Sakai *et al.* 1998] The specific feature of the deposited particles in PLD is that the energy of ion and neutral atom are very different, such as the ions have energies of several keV and the neutral atoms have those of several eV.

3. 1. 2. Thin Films Prepared by PLD

PLD of thin films have several advantages like the followings in comparison with physical vapor deposition (PVD) method besides sputtering and vacuum deposition ablation. In this study, some of the advantages had been practically used. A material of high-melting point (m.p.) is easy to be deposited, since a laser beam of high energy density is used. An alternative deposition of multiple materials is possible, since the ablation is pulsed process that can be controlled digitally. In this study, mixture thin films of the carbon - nickel - indium are deposited by the alternative laser ablation of multiple materials, carbon (m.p. > 3500 °C), nickel (m.p. = 1455 °C) and indium (m.p. = 156.6 °C). The pollution in the deposition chamber is not substantial, since only the laser beam was introduced through the window from the outside to irradiate and heat only very small area. The thickness of the film is easy to control, since one laser pulse deposits only 0.002 nm. In the meantime, there are some disadvantages like the adhesion of debris. There are some techniques that remove the debris to prepared flat surface. [Chen 1994, György *et al.* 2002] However no effort to remove the debris have done in this study, since it does not seem to disturb the growth of nanowhiskers.

3. 2. Features of Alloy Whisker

Nanowhiskers grown in this study consist of nickel (Ni) - indium (In) alloy. Here this binary system is considered.

Fig. 15 shows the phase diagram of nickel - indium alloys. [Hellner 1950] However, it is not always accurate, since the different phase diagrams are otherwise being announced. [Hellner and Laves 1947, Baranova and Pinsker 1965]

The composition ratio of the grown nanowhisker was observed by EDS, and it was Ni : In = 2 : 3. It is well known that carbon (C) and indium (In) hardly forms a compound. The melting point of the nickel carbide (Ni-C) is 1326.5 °C, which is much higher than the annealing temperature. From these facts, it is needless to consider the carbonic chemical compound. Here, it seemed that nickel (Ni) and indium (In) reacted and form the Ni_2In_3 alloy. Fig. 16 shows the part of consideration in the phase diagram in Fig. 15. The growth of the alloy is explained basing on this phase diagram. In the prepared graphite thin film, indium and nickel deposited as dispersing sufficiently. The experimental results show the nanowhiskers did not grow when the substrate temperature was 200, 300 and 400 °C, but RT and 100 °C. That is to say, the dispersed indiums melt and make liquid phase at the substrate temperature of over the melting point of indium (156.6 °C); moreover they come together and make droplets afterwards. It was already mentioned in the experimental results, while it is discussed with the phase diagram (Fig. 16).

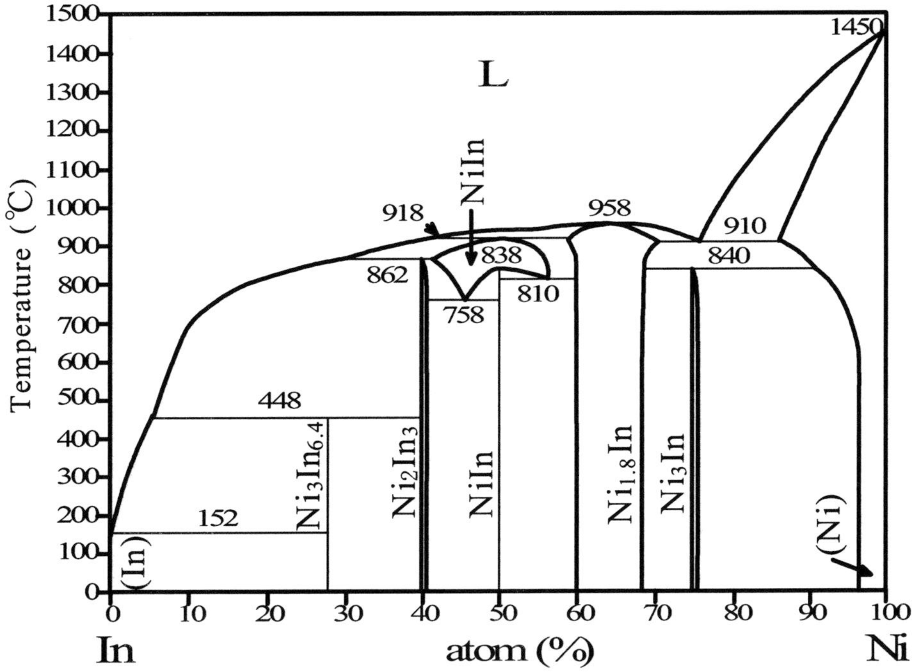

Figure 15. Phase diagram of Ni-In alloy. [Hellner 1950].

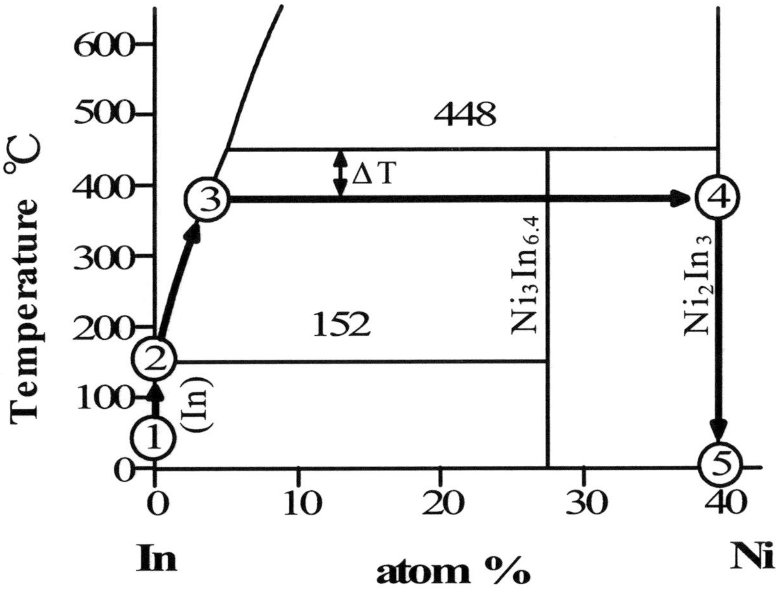

Figure 16. Related part of the phase diagram of Ni-In alloy. In the solution phase along the line (3) -(4), it is a supercooling state. The differential temperature ΔT between the peritectic and actually solution phase traced is the supercooling degree, which produces the driving force of the crystal growth.

In the phase diagram of Fig. 16, the sufficiently dispersed indium is located (1) in the as-deposited condition. Until the temperature reaches to the melting point, it traces the line (1) - (2). As the temperature rises, the liquid indium traced on the liquidus curve (2) - (3) with taking in nickel. If the quantity of nickel is not sufficient, it deviates from the liquidus curve (2) - (3) and the liquid of the indium separates. When the temperature increases along the liquidus curve to (3) and the solid phase starts to grow as a supercooled state, it takes in nickel along the solidus curve (3) - (4) as a solution phase. The solution takes in nickel and begins to generate crystals of Ni_2In_3 at (4), thus the quantity of nickel decreases. The separation from the solution phase precedes the indium rich direction along the solidus curve (4) - (3). When the separation perfectly advances, the solution will be separated into the solid phase of nanowhisker Ni_2In_3 and the liquid phase of indium (In). By cooling, the liquid phase of indium (In) is solidified along (3) - (2) - (1), and it becomes a droplet state. The nanowhiskers Ni_2In_3 cool down along the solidus line (4) - (5). In this way, only Ni_2In_3 and In are precipitated on the surface of the thin film.

The supercooling is considered as a driving force of the crystal growth. That is to say, the differential temperature ΔT between the peritectic temperature in Fig. 16 and actually solution phase traced seems to be the supercooling degree. For normal line growth rate R of the crystal growth is in proportion to the differential temperature ΔT, which is also means the supercooling degree. When ΔT is fixed, the normal line growth rate R of the crystal growth is also constant. The similar crystal keeps growing with supply of the solution phase. This sufficiently satisfies the growth condition of whiskers with large aspect ratio. Here, Table 3 summarizes the properties of Ni_2In_3 and $Ni_3In_{6.4}$ for comparisons. [Hellner 1950, Baranova and Pinsker 1965] Fig. 17 shows the crystalline structure of Ni_2In_3.

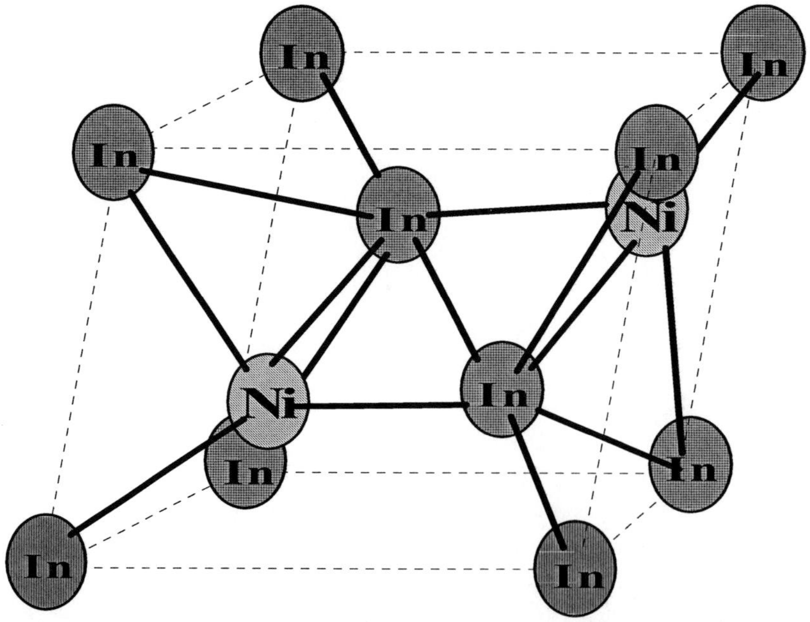

Figure 17. Crystalline structure of Ni_2In_3

Table 3. Properties of Ni_2In_3 and $Ni_3In_{6.4}$

	Ni_2In_3	$Ni_3In_{6.4}$
Standardized Data		
Structure Type	Ni_2Al_3	Ru_3Sn_7
Pearson symbol	hP5	cI40
Space group	P-3m1	Im-3m
Space group No.	164	229
Z	1	4
Cell Parameters		
a (nm)	0.4387	0.918
b (nm)	0.4387	0.918
c (nm)	0.5295	0.918
a (°)	90	90
b (°)	90	90
g (°)	120	90
a/b	1.000	1.000
b/c	0.829	1.000
c/a	1.207	1.000
Cell volume (nm^3)	0.08825	0.77362
Density (Mg m^{-3})	8.69	7.82

With the analysis by the TEM micrograph (Fig. 8), it was observed that the nanowhiskers had grown on a graphite thin film. The TEM image shows the graphite film is seen to be a patchy pattern and a sponge-like condition in the inside, after creeping out of the simultaneously deposited nickel and indium. The diameter of the nanowhisker is about 20 nm; furthermore the thickness of the film is also about 20 nm. It is not reasonable that the cylindrical whisker filler with contents, therefore the nickel and indium sources are not supplied from the envelopment considered from the thickness and the aspect ratio. Furthermore, there is a double structure in the edge of the whisker in the TEM image. The fact above mentioned shows that the whisker is hollow structured.

3. 3. Growth Model of Alloy Nanowhisker

On the growth of whiskers, many experiments have been carrying out, and many theories have been explained. The most popular mechanism is vapor-liquid-solid (VLS) method, in which a solid whisker grows from an impurity seed that becomes supersaturated liquid with material supplied from the vapor. [Wagner 1970, Kaneko 1993] Fig. 18 shows VLS method. However, other model must be searched to explain the present alloy nanowhisker, since there is no impurity seed in the tip either and the whiskers were formed in both vacuum and atmosphere.

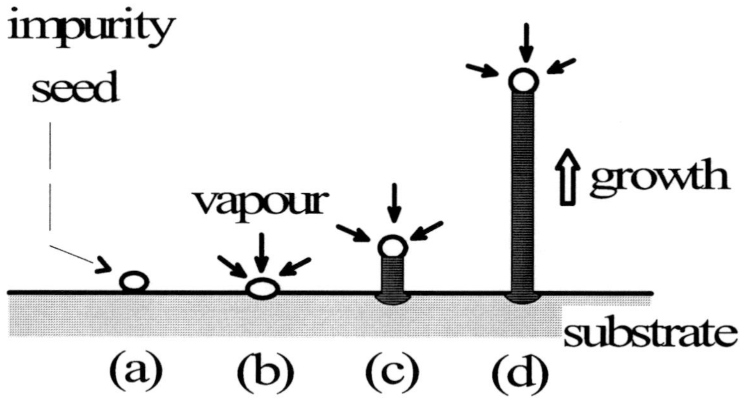

Figure 18. Schematic drawing of vapor-liquid-solid (VLS) method for whisker growth. An impurity seed melts (a) and forms a crater (b). The seed absorbs the vapor and become supersaturated, it fellows the nucleation and initial growth of a whisker (c). The whisker continued to grow (d).

There is an example of producing nanowire of bismuth (Bi) by the residual stress of evaporated thin film. In this method, it is possible that the nanowire with very large aspect ratio can be produced. This method might be applied to the alloy whose melting point is low. [Cheng *et al.* 2002]

The driving force of the present alloy nanowhisker is not residual stress in the prepared thin film but crystallization from supercooled state as discussed. The residual stress in the prepared film forces to transfer the liquid indium grains during the annealing. Nevertheless the growth of nanowhiskers is not performed only by this transfers produced by residual stress of physical energy. A supersaturation of chemical energy is necessary for crystal growth.

Some similar mechanisms related to this phenomenon were studied formerly in the growth of NaCl and KCl from alkali halide aqueous solution. They grew from inside of the semi permeable membrane by creeping of the film due to capillary forces. [Shichiri 1974, Shichiri and Kato 1968, Amelinckx 1959] In these methods, the growths in both of part and tip minute of the root of the whisker were observed.

In the method of this study, the tip of the whisker does not contain either the seed particle anywhere. Besides, the liquid alloy dose not become a large liquid metal inside of the graphite thin film, since the thickness of the film is only 20 nm in which there are not enough space. So therefore the alloy of the supercooled state creeped out by residual stress of the porous graphite thin film on the surface used the axial capillarity force from the solution phase. Furthermore it is possible to explain the growth of the alloy nanowhiskers based on a spiral dislocation with the axial capillarity force. Fig. 19 shows the mechanism of the nanowhisker growth by the capillarity. The porosity of graphite thin film was formed by scattering the atomic particles of nickel and indium in the inside of the graphite thin film. By annealing, the indium particles become liquid at 156.6 °C, and the nickel particles are taken in and become $Ni_3In_{6.4}$ or Ni_2In_3 under 400 °C as a supercooled state. The alloy seems to be a supersaturation state at this time, since melting point of the alloy is 410 °C. It is creeped out by the capillarity in the screw axis direction, when the liquid metal exhausts it in a film, and it crystallizes rising at the kink points. In the reason, the whisker of rod like, roll paper like and pipe like state are formed. The size of holes in the graphite film, the conditions such as the quantity of liquid alloy, the residual stress of the film decide the forms of whiskers. In this study, the moderate nanowhisker grows, when the diameters of the holes are almost 5 ~ 50 nm. However, the nanowires do not grow but island states developed in the case when it is more greatly over 100 nm. It is possible to explain the composition of observed nanowhisker growths with the model, when the consideration of experimental results and the above-mentioned proposals are taken into consideration.

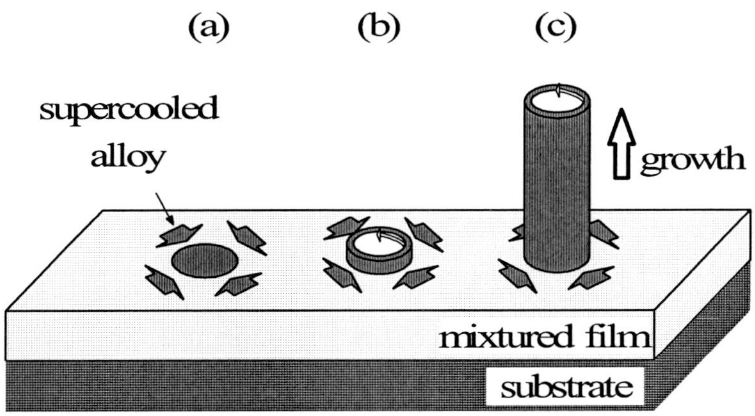

Figure 19. Schematic drawing of the growth model of alloy nanowhisker. Supercooled alloy form an initial state of whisker (a). By the continuing supply of supercooled liquid alloy, it is creeped out by the capillarity in the screw axis direction, and it crystallizes rising at the kink points. (b) The whisker continued to grow (c).

4. APPLICATIONS

In this section, some novel applications of the present alloy nanowhisker are proposed. The related ideas have already been presented in many introductory books of the nanotechnology. Here, they are mentioned considering the potentials of sufficiently utilizing the advantage of the alloy and the hollow structure. It must be noticed that the references in this section are only some representatives, since there are numerous references in nanotechnology. [Nalwa 2000, Timp 1999, Sakaki 2004, Nogi 2003]

To begin with, it must be known which combinations of metals are possible for the growth of the alloy nanowhisker. The alloy nanowhiskers are grown in respect of the carbon film of multiple metals dispersed by laser evaporation and annealing. This technique is able to apply to other alloys, while it is possible to produce the alloy nanowhisker of various compositions including the incongruent melting type chemical compound. The alloy with possibility of growth of nanowhisker consists like the following; for example, Ag-In, Ag-Nd, Al-Ba, Al-Ca, Au-Ga, Mg-Au, Ba-In, Bi-In, etc.

For the size and expected electrical conduction, much new technology will develop by the alloy nanowhiskers. For example, they are various fields of organic electroluminescence from the quantum size effect, spitronics from the confinement effect, medical application from the miniaturization and scanning probe microscopy from the electrical conduction property.

By the electric conductivity of the alloy nanowhisker, it is possible to develop a miniaturization of existing electronic devices for the smallness. Furthermore, completely new types of devices are possible to be developed with this technology, since a nature that a single electron can move only in uni-direction becomes also possible in the tiny nanowhisker that can be re-formed as a craft paper. If the size becomes within one nanometer square, it is possible to produce a quantum dot in which a single electron can be included for each dot. The quantum dot can be used not only for a quantum bit, but also for a preparation of optics switches for transmitting data by miniature semiconductor laser and photo detector, the telecommunication network of the future, etc. [Timp *et al.* 1999] The technique of forming quantum dots by cutting down from the alloy nanowhisker seems easy. On the other hand, if it were possible to control the size, form, structure of the alloy nanowhisker in the growth, it would be able to freely design handling quality of optical elements such as laser structures. [Huang *et al.* 2001]

Since the alloy nanowhisker have hollow structures, there is a possibility to form "quantum confined atoms" (QCA). QCA is an atom or an ion (atom with electric charge) trapped within a "nanocrystal cage", that is a tiny cage made from semiconductor atoms. For carbonic allotropes, "endofullerenes", which contain other atoms in fullerenes, are the representative of QCA. [Akasaka and Nagase 2002] In the conventional semiconductor technology, the properties of semiconductor are decided by the trapped atom, however in QCA, the atoms of the cage modify the properties of the confining atom. QCA has a possibility of becoming a basis of the new semiconductor device, such as memory or storage materials for data in ultra-high-density magnetic recording system and various equipments in portable audio-visual players and mobile phones. [Sakaki 1999] In addition, if an atom can be trapped in the nanowhisker, it is possible to apply a new light emitter, which can use as light, displays, biomarkers, etc. It also can be used for photo magnetism (magnet and sensor), manufacturing of laser for optical fiber communication equipment and ultimate photon quantum computers. [Bhargava 2004, Govorov 2004]

The alloy nanowhisker is considered for the application to biotechnology and medicament in utilizing the component and shape. For example, a drug delivery is one of the solutions. A life body takes in the necessary material from digestive organs and roots through blood vessels at each part in order to maintain the life. In the each place, there are immune function and selective function in order to sort out noxious material and useful material. Such function is performed in nanoscale. The alloy nanowhisker may be possible to send only the necessary things to the necessary place, since the component and size are controllable. The hollow alloy nanowhisker is made to be a capsule, the medicine is put in, and it is sent to the lesions. The capsule can open by electromagnetic stimulation or ultrasonic waves from the outside at the objective place, thus the medicine is released. It can be applied for the gene therapy. [Verma and Somia 1997] It is very interesting to imagine that a procaryote in a metallic cage is delivering to a lesion. By such drag delivery systems, the medicament may advance. [Chien 1982, Sakaki 2004] It is also considered that some nanomachines that apply to the biotechnology can be produced from the negligible shape. The variety of alloy nanowhiskers in the both shapes and properties may provide the parts for sub-micrometer-sized robot.

The alloy nanowhiskers have potentials to decrease the film thickness of the electrical conduction layer of the transistor and the high-speed gate electrodes to several nanometers. A paper-like material that consists of hexagonal close-packed bundles of the alloys might be fabricated using similar methods like carbon nanotube. [Endo *et al.* 2005] It follows that there are possibilities for the next generation electromagnetic wave of the region over 10 GHz, which lies from microwave to infrared light in the electromagnetic spectrum. For example, the electromagnetic wave of 60 GHz band is expected to apply for the collision preventions for automobile, the indoor wireless LAN and the intelligent transportation system (ITS), since the propagated distance in the air is short. Furthermore the electromagnetic wave of THz band, which is called T-ray and has a wavelength of about one-tenth of a millimeter, is an ideal candidate for medical imaging, since it is too large to scatter randomly off cell walls but enough small to provide images with useful resolution. The single photon energy of T-rays is not enough high to separate the electron from the intracellular atom; subsequently it is safer than the X-ray for no anxiety of causing the mutation of the cell. [Zandonella 2003]

The applications cited above are not the prospective examples. The more concrete applications will be suggested as the physical, chemical and mechanical properties of the nanowhiskers are clarified.

5. Conclusion

The researches of the alloy nanowhisker have not performed yet, thus the details of the growth mechanisms and optimum conditions, fundamental physical property, etc., are unsolved. In this study, some additional experiments, such as AFM observations, Rutherford backscattering spectroscopy (RBS), in-situ optical microscope observation, have been tried to reveal them, but they have not come to fruition.

When I was a primary school boy, the almighty material that came out in science fictions was "superalloy". The superalloy was also used for the main material of inter-galaxy rockets, space robots and exploration ships. However, the alloy does not become a main material in present nanotechnology. If the superalloy described in this research shoulders future science and technology, the arrival of near future science fiction may be foreboded.

Acknowledgements

I would like to acknowledge the collaborations with K. Hojou, S. Jitsukawa, H. Koshikawa, V. Lavrentiev, C. Lin, A. Naito, H. Naramoto, K. Narumi, Y. Saitoh, M. Sataka, T. Sawai, A. Takeyama, S. Yamamoto, Y. Xu and Z. Yoshida at JAERI. A special debt of gratitude goes to my former teachers: S. Ohtani, H. Takuma and K. Shimoda.

References

Akasaka T., Nagase S., (2002) *Ed.* "Endofullerenes", Kluwer Academic, Dordrecht.

Amelinckx S., (1959) "Mechanism for the Growth of Whiskers from Solution on an Impervious Substrate" *J. Chem. Phys.* 31, 1687-1688.

Appell D., (2002), "Wired for success", *Nature*, 419, 553-555.

Ashcroft N W., Mermin N D., (1976) "Solid State Physics" Saunders College, Philadelphia.

Ball P., (2001), "Roll up for the revolution", *Nature*, 414, 142-144.

Baranova R.V., Pinsker Z.G.,(1965) "Electron-diffraction study of the structure of the g phase in the Ni-In system" *Soviet Physics-crystallography*, translated from Kristallografiya, 10, 523-528. through *Pauling file* (http://crystdb.nims.go.jp/index.html).

Bashkin S., Storner Jr. J O., "Atomic energy-level and grotorian diagrams" vol. I., North-Holland pub. comp. Amsterdam (1978a).

Bashkin S., Storner Jr. J O., "Atomic energy-level and grotorian diagrams" vol. II., North-Holland pub. comp. Amsterdam (1978b).

Belitskus D., (1993) "Fiber and Whisker Reinforced Ceramics for Structural Applications", Marcel Dekker, New York.

Bhargava R. N. (2004), "Quantum-con"ned atoms" a: novel luminescent centers for future II-VI devices" *J. Crystal Growth* 214-215, 926-930.

Björk M. T., Ohlsson B. J., Sass T., Persson A. I., Thelander C., Magnusson M. H., Deppert K., Wallenberg L. R., Samuelson L., (2002) "One-dimensional heterostructures in semiconductor nanowhiskers", *Appl.Phys. Lett.* 80, 1058-1060.

Chen L-C., (1994), "Particulates Generated By Pulsed Laser Ablation" in "*Pulsed Laser deposition of thin films*" (D. B. Chrisey and G. K. Huber, eds), John Willy and sons inc, New York, Chapter 6.

Chen F F., (1974), "*Introduction to Plasma Physics*", Plenum Press, New York.

Cheng Y-T., Weiner A. M., Wong C. A., Balogh M. P. and Lukitsch M. J., (2002) "Stress-induced growth of bismuth nanowire", *Appl. Phys. Lett.* 81, 3248-3250.

Chien Y. W., (1982) "*Novel drug delivery systems*" Marcel Dekker, New York.

Dresselhaus M.S., Dresselhaus G. and Eklund P.C., (1996) "*Science of Fulerenes and Carbon Nanotubes*", Academic Press, London.

Endo M., Muramatsu H., Hayashi T., Kim Y. A., Terrones M., Dresselhaus M. S. (2005), "Buckypaper' from coaxial nanotubes", *Nature*, 433, 476.

Endo M., Iijima S., Dresselhaus M. S., (1996) *Ed.* "*Carbon nanotubes*", Elsevier, Oxford.

Ferrari A C., Robertson J., (2000) "Interpretation of Raman spectra of disordered and amorphous carbon", *Phys Rev. B*, 61: 14 095-107.

Fujiki Y., Mitomo M., (1993), "*Whisker*" Sangyo tosho, Tokyo. [in Japanese]

Geohegan D. B., (1994), "Diagnostics and Characteristics of Laser-Produced Plasmas" in "*Pulsed Laser deposition of thin films*" (D. B. Chrisey and G. K. Huber, eds), John Willy and sons inc, New York, Chapter 5.

Govorov A. O. (2004), "Optical probing of the spin state of a single magnetic impurity in a self-assembled quantum dot", *Phys. Rev. B* 70, 035321-035326.

Gudiksen M. S., Lauhon L. J., Wang J., Smith D. C., M. Lieber C., (2002) "Growth of nanowire superlattice structures for nanoscale photonics and electronics", *Nature* 415, 617-620.

György E., Mihailescu I.N., Kompitsas M., Giannoudakos A., (2002) "Particulates-free Ta thin films obtained by pulsed laser deposition: the role of a second laser in the laser-induced plasma heating" *Applied Surface Science* 195, 270–276.

Hellner E. (1950) "Das System Nickel-Indium" *Zeitschrift fuer Metallkunde* 41, 401-406. [in German] through *Pauling file* (http://crystdb.nims.go.jp/index.html).

Hellner E., Laves F., (1947) "Kristallchemie des In und Ga in Legierungen mit einigen Übergangselementen (Ni, Pd, Pt, Cu, Ag und Au)" *Zeitschrift fuer Naturforschung, Teil*

A: Astrophysik, Physik und Physikalische Chemie 2, 177-183. [in German] through *Pauling file* (http://crystdb.nims.go.jp/index.html).

Hotta K. *et al,* (1999) "*Laser Ablation and Applications*" (The Institute of Electrical Engineers of Japan, eds), Coronasha, Tokyo. [in Japanese].

Hughes R H., Anderson R J., (1989) "Laser Ion Sources" in ""*The Physics and Technology of Ion Sources*" (BROWN, I.G., eds) John Willy and sons inc, New York, Chapter 14.

Huang M H, Mao S., Feick H., Yan H., Wu Y., Kind H.,Weber E.,Russo R., Yang P., (2001) "Room-Temperature Ultraviolet Nanowire Nanolasers" *Science* 292, 1897-1899.

Iijima S., (1991), Helical microtubules of graphitic carbon", *Nature*, 354, 56-58.

Kaneko T., (1993) "*Hige Kessho*", Kyoritsu shuppan, Tokyo. [in Japanese].

Kelly R., Miotello A., (1994), "Particulates Generated By Pulsed Laser Ablation" in "*Pulsed Laser deposition of thin films*" (D. B. Chrisey and G. K. Huber, eds), John Willy and sons inc, New York, Chapter 3.

Kitazawa S., Abe H., Yamamoto S., (2005), "Formation of nanostructured solid-state carbon particles by laser ablation of graphite in isopropyl alcohol", *J Phys. Chem. Solids.*, 66/2-4, 555-559.

Kitazawa S., (2004) "In-situ Optical Spectroscopy of Ablation Plume for Preparations of Nanostructured TiO_2 Thin Films by Pulsed Laser Deposition", *Jpn. J. Appl. Phys.* 43/9A, 6335-6341.

Kitazawa S., Naito A., LIN C. and Naramoto H., (2004a) "Growth of Ni-In Alloy nano-whiskers", *J. Crystal Growth* 267, 336-339.

Kitazawa S., Choi Y., Yamamoto S., (2004b) "In-situ optical spectroscopy of PLD of nano-structured TiO_2" *Vacuum*, 74/3-4, 637-642.

Kitazawa S., Putra T. Y. S. P., Sakai S., Narumi K., Naramoto H., Yamamoto S., Chiba A., (2003a) "Laser detection of surface acoustic waves as a method of measuring an Ar ion beam modification of carbon thin film" *Nuclear Instruments and Methodes in Physics Research B* 206, 952-955.

Kitazawa S., Yamamoto S., (2003b) "Optical Spectroscopy of Plasma Plume Induced by Pulsed Laser Deposition of TiO_2 Thin Films" *Transactions of Materials Research Society of Japan*, 28 [4], 1133-1136.

Kitazawa S., Okada K., Nakamura T., Ohtani S., Takuma H., (1993) "Trapping of Multiply Charged Ta Ions produced by Laser Ablation in an Ion Trap" *Jpn. J. Appl. Phys.* 32, 5147-5148.

Levitt A. P., (1970) *Ed.* "*Whisker Technology*", Wiley, New York.

Lin C., Naramoto H., Xu Y H., Kitazawa S., Narumi K., Sakai S., (2003) "Formation of nano-sized pinholes array in thin Ni film on MgO(100) substrate" *Thin Solid Films* 443, 28-32.

Lin C., Xu Y H., Naramoto H., Wei P., Kitazawa S., Narumi K., (2002) "Morphology evolution of thin Ni film on MgO(100) substrate" *J. Phys. D: Appl. Phys.* 35, 1864-1866.

Martin W. C., *et. al.*, (1999) "*NIST Atomic Spectroscopic Database* Ver. 1.3" (http://physics.nist.gov).

Murakami K., (1999), in "*Laser Ablation and Applications*" (The Institute of Electrical Engineers of Japan, eds), Coronasha, Tokyo, p11. [in Japanese].

Nalwa H. S., (2000) Ed., "Handbook of Nanostructured Materials and Nanotechnology; vol. 1 Synthesis and Processing", Academic Press, San Diego.

Nogi K., (2003) *Ed.,* "*Nanoparticle Technology*", Nikkan-kogyo shinbun, Tokyo. [in Japanese]

Okada T., (1999), in "*Laser Ablation and Applications*" (The Institute of Electrical Engineers of Japan, eds), Coronasha, Tokyo, p260. [in Japanese].

Saenger K. L., (1994), "Angular Distribution of Ablated Material" in "*Pulsed Laser deposition of thin films*" (D. B. Chrisey and G. K. Huber, eds), John Willy and sons inc, New York, Chapter 7.

Sakai H., Keitoku S., Ezumi H., (1998), "Contribution of High-Velocity Particles to Stress of Thin Film Produced by Laser Ablation", *Jpn. J. Appl. Phys.* 37, 4149-4153.

Sakaki H., (2004) "*Nanotechnology*", Kanki Shuppan, Tokyo. [in Japanese].

Sakaki H., (1999) "Fabrication of Atomically Controlled Nanostructures and Their Device Application" in "*Nanotechnology*", (Timp G., ed.), Springer, New York, Chapter 5.

Service R. F.,(2002) "Nanowire Fabricators Earn Their Stripes", *Science* 295, 946-947.

Shichiri T. (1974) "The growth mechanism of NaCl whiskers" *J. Crystal Growth* 24/25, 350-353.

Shichiri T., Kato N., (1968) "Regrowth of NaCl whiskers from pure and poisoned solutions", *J. Crystal Growth* 3, 4, 384-390.

Singh R. K., Narayan J., (1990) "Pulsed-laser evaporation technique for deposition of thin films: Physics and theoretical model", *Phys. Rev.* B 41, 8843-8859.

Tanaka K., (2002), "The Origin of Macromolecule Ionization by Laser Irradiation", Nobel Lecture, December 8, 2002, at Aula Magna, Stockholm University.

Timp G., *Ed.*, (1999) "*Nanotechnology*", Springer, New York.

Timp G., Howard R. E., Mankiewich P. M., (1999) "Nano-electronics for Advanced Computation and Communication" in "*Nanotechnology*", (Timp G., ed.), Springer, New York, Chapter 2.

Verma I M., Somia N., (1997), "Gene therapy - promises, problems and prospects" *Nature* 389, 239 - 242.

Villars P., *et al.* (2001) *Eds.* "Pauling File: Basic Database for Crystal Structures" (http://crystdb.nims.go.jp/index.html)

Wagner R. S., (1970) "VLS Mechanism of Crystal Growth" in "*Whisker Technology*", (Levitt A. P., ed.) Wiley, New York, Chapter 3.

Williams D B., Carter C B., (1996) "*Transmission Electron Microscopy*", Plenum Press, New York.

Yamamoto M., Murayama S., (1995) "*Plasma no bunkou-keisoku*", Japan Scientific Societies Press, Tokyo. [in Japanese]

Zandonella C., (2003), "T-ray specs", *Nature*, 424, 721-722.

In: New Topics in Crystal Growth Research
Editor: George V. Karas, pp. 153-177 ISBN 1-59454-837-4

Chapter 4

THE GROWTH OF LOW RESISTANCE METAL SILICIDE FILMS BY SOLID STATE REACTION

J. Y. Feng**, *Y. He, J. F. Liu and R. N. Wang
Department of Materials Science and Engineering
Tsinghua University, Beijing 100084, CHINA

ABSTRACT

In recent years, low resistance metal silicides have been generally used in CMOS technology to decrease the sheet resistance of highly doped regions. In this article, we report the growth of three generations contact materials, i.e. C54-$TiSi_2$, $CoSi_2$ and NiSi, by solid state reaction in advanced CMOS technology. The research focuses on three aspects: the formation of C54-$TiSi_2$; the epitaxial growth of $CoSi_2$ on Si(100) substrates and the thermal stability of NiSi films.

In Ti-Si system, a thin layer of Ta was introduced between Ti films and Si substrates in order to promote the formation of C54-$TiSi_2$. After rapid thermal annealing (**RTA**) at 580℃, the single C40-$TiSi_2$ phase was found in the sample with Ta interlayer, and C54-$TiSi_2$ phase was formed directly, bypassing the C49-$TiSi_2$ when annealed at higher temperature (650℃). The formation temperature of C54-$TiSi_2$ is lowered by 100℃ compared with the sample without the Ta layer.

In Co-Si system, highly textured Co films were obtained by heating the Si substrates to an adequate temperature during the deposition of cobalt. The direct solid phase reaction between textured Co films and Si(100) substrates at high temperatures produced epitaxial $CoSi_2$ with good quality. Rutherford backscattering/channeling spectrometry and high resolution transmission electron microscopy were performed to evaluate the epitaxial quality of the $CoSi_2$ films. A good channeling χ_{min} value of 20% was measured for the $CoSi_2$ films formed after rapid thermal annealing at 850℃, which is comparable to that of the $CoSi_2$ films obtained by Ti-interlayer mediated epitaxy **(TIME)**. Similar with **TIME**, the intermediate phase of Co_2Si was skipped in this case. The concept of "thermodynamic barrier" has been employed to explain the improvement in the epitaxy of $CoSi_2$ films compared with the reaction between amorphous Co films and Si substrates.

* Corresponding authour: Tel: 86-10-62771575; Fax: 86-10-62771160; email: fengjy@mail.tsinghua.edu.cn

In Ni-Si system, a thin Pt interlayer was used to raise the phase transition temperature by about 100℃ and thereby greatly improve the thermal stability of NiSi films. It was also found that the thermal stability of NiSi films improves with the decrease of the film thickness in Ni/Pt/Si system, implying a promising future for the application of NiSi films in shallow junction contacts. A semi-quantitative analysis was carried out to clarify the effect of Pt on the activation energy for nucleation and critical nucleus size of $NiSi_2$, which explains the above the experimental results.

Key Words: $TiSi_2$; Phase transition; Crystalline orientation; $CoSi_2$; Epitaxy; Texture(Preferred orientation); Nucleation; NiSi; Thermal stability; Interface eneygy

PACS: 68.35.-p; 68.55.-a; 68.60.Dv; 73.50.-h

1. Introduction

Low resistance metal silicide is an important part of integrated circuit technology. In order to reduce the sheet resistance of highly doped regions in CMOS (complementary metal oxide silicon), metal silicide thin films have been introduced in CMOS technology in the past years [1, 2]. In this article, we will focus on the use of low resistance metal silicides (i.e. $TiSi_2$, $CoSi_2$ and NiSi) in advanced CMOS technology.

The metal silicides are formed by a solid state reaction between metal and silicon. The phase formation kinetics of the reaction between metal films and Si has been investigated since the late 1960s [3]. In recent years, the research has been focusing on reducing the sheet resistance and controlling Si consumption during thin films formation in confined areas [2]. In this article, we will report the performance of three generations low-resistance metal silicide thin films (i.e. C54-$TiSi_2$, $CoSi_2$ and NiSi) in details.

2. Investigation of Three Generations Contact Materials

2.1. Formation of C54-$TiSi_2$ at Low Temperature

$TiSi_2$ is widely used as contacts and interconnects for the metallization of gates and source/drains in the ultra-large scale integration (**ULSI**) devices and circuits. However, the high-resistance C49-$TiSi_2$ forms on heating Ti on Si above 500°C, and additional heating above 700°C is needed to transform C49 into the low-resistance C54-$TiSi_2$ phase [4, 5]. This transformation becomes more difficult in submicron Si lines, because of the low density of nucleation sites [6, 7]. In recent papers [8-10], it was shown that the presence of Mo, Ta or Nb impurities at the metal-silicon interface lowers the formation temperature of the C54 phase with respect to the pure Ti-Si system. In a first attempt this reduction was attributed to the suppression of the intermediate C49 phase [11]. Later measurements by high-resolution transmission electron microscopy (**HRTEM**) indicated the formation of a ternary compound C40 (Ti, Ta)Si_2 at the Ti-Si interface that acts as a seed for the successive growth of the C40 $TiSi_2$ phase [12]. Then the C40 phase allows the direct formation of C54 phase as a template by blocking the formation of C49 phase.

In fact, this new intermediate C40 $TiSi_2$ phase has been detected in the work of F. La Via and his co-workers [13] and they calculated the lattice parameters for this new phase theoretically by using the local density approximation (**LDA**), the generalized gradient approximation (**GGA**), *etc.* K. Li *et al* [14] formed the C40 $TiSi_2$ phase by pulsed-laser annealing, and identified that the C40-$TiSi_2$ has a hexagonal structure with space group $P6_222$ and lattice parameters $a = 0.471nm$ and $c = 0.653nm$ by convergent beam electron diffraction (**CBED**) and **CBED** pattern simulation; in the same way, T. Yu [15] formed the C40-$TiSi_2$ and studied in detail by glancing-angle X-ray diffraction (**GAXRD**).

Here, we introduced a Ta interlayer between Ti films and Si substrates to suppress the formation of C49-$TiSi_2$. A 2nm thin layer of Ta and a 50nm thick layer of Ti were sequentially e-gun evaporated on Si(001) substrates forming a bi-layered system, as we called Ti/Ta/Si samples. To make a comparison, 50nm-thick Ti films without Ta interlayer were also deposited on Si(001) substrates under the same experimental conditions, called Ti/Si samples. Rapid thermal annealing (**RTA**) of the samples was carried out in a high vacuum of 2×10^{-6}Torr at temperatures ranging from 500 to 800℃ for 5 min. The pure C40-$TiSi_2$ films were found when the samples with the Ta interlayer were annealed at 580℃, and the lattice parameters of C40-$TiSi_2$, were obtained according to the **XRD** data and compared with others' work. The effects of the pure C40-$TiSi_2$ films on the formation C54-$TiSi_2$ were given [16].

The **XRD** patterns of Ti/Si samples annealed at different temperatures are shown in Fig. 1. It can be seen that after annealing at 530℃, the preferred (102) orientation of Ti_5Si_3 phase is formed [Fig.1 (a)]. It is shown as Fig.1 (b) that when the Ti/Si samples are annealed at 650℃, the Ti_5Si_3 phase disappears with only C49 phase present. The C54-$TiSi_2$ orientations don't appear, and become dominant until the Ti/Si samples are annealed at 750℃ [Fig.1 (c)], indicating that the transformation from C49 to C54 phase is completed at this temperature. For Ti/Ta/Si samples, Ti_5Si_3 is also the first phase when annealed at 530℃. The un-reacted Ti is oxidized with the Ti oxide diffraction peaks in all the Ti/Ta/Si samples [Fig.2 (a)-(c)] probably because the Ta interlayer blocks the diffusion of Ti and Si atoms and limit the silicidation reaction, resulting in some Ti left and oxidized. When the annealing temperature is raised to 580℃, the single C40-$TiSi_2$ phase with the preferred orientation C40 (003) is formed without any other peaks corresponding to Ti silicides [Fig.2 (b)]. The direct transition from the C40-$TiSi_2$ to C54-$TiSi_2$ phase is completed when the Ti/Ta/Si samples are annealed at 650℃ [Fig.2(c)], bypassing the C49-$TiSi_2$ phase. From these data, it can be concluded that the C54 $TiSi_2$ films are formed completely at 650℃ for Ti/Ta/Si samples, while this formation doesn't complete until the annealing temperature rises to 750℃ for Ti/Si samples. It is consistent with the results of the measurement of the sheet resistance by the 4-point probe for Ti/Si and Ti/Ta/Si samples annealed at the same temperature ranging from 530 to 780℃. For Ti/Ta/Si samples, **XRD** results (not shown in this article) indicate that the Ti oxide on the surface of the samples has been removed after immersing the wafers sequentially in H_2O_2:H_2SO_4 (1:4), NH_4OH:H_2O_2:H_2O (1:1:5), and the de-ionized water. As shown in Fig.3, it can be seen that when both kinds of samples are annealed at the same temperature, the sheet resistance of Ti/Si samples is usually higher than that of Ti/Ta/Si samples; the sheet resistance for Ti/Ta/Si samples falls into a low level (less than 10Ω/sqr.) between 600 and 700℃, indicating the formation of the low-resistance C54-$TiSi_2$, but this doesn't happen until the temperature rise to over 700℃ for Ti/Si samples. Thus it can be safely concluded that the addition of the thin Ta interlayer lowers the C54-$TiSi_2$ phase formation temperature by nearly 100℃.

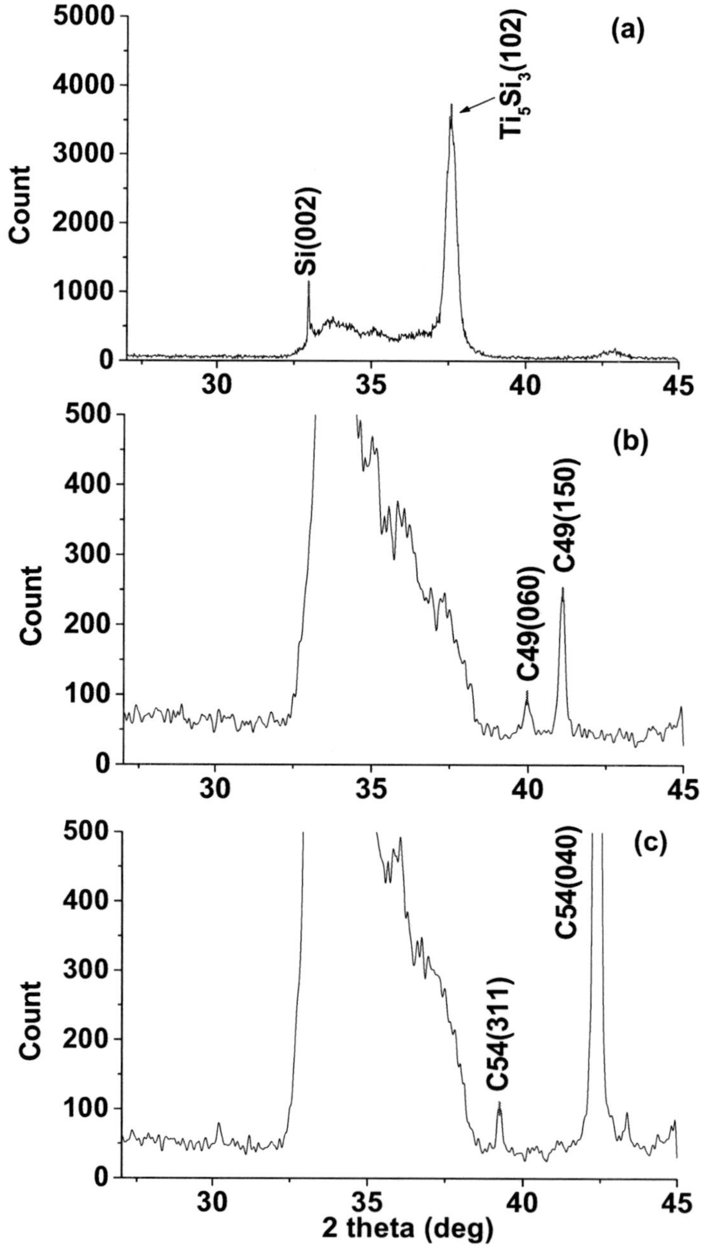

Figure 1. XRD patterns of the Ti/Si samples annealed at 530℃ (a), 650℃ (b), and 750℃ (c) respectively for 5 min

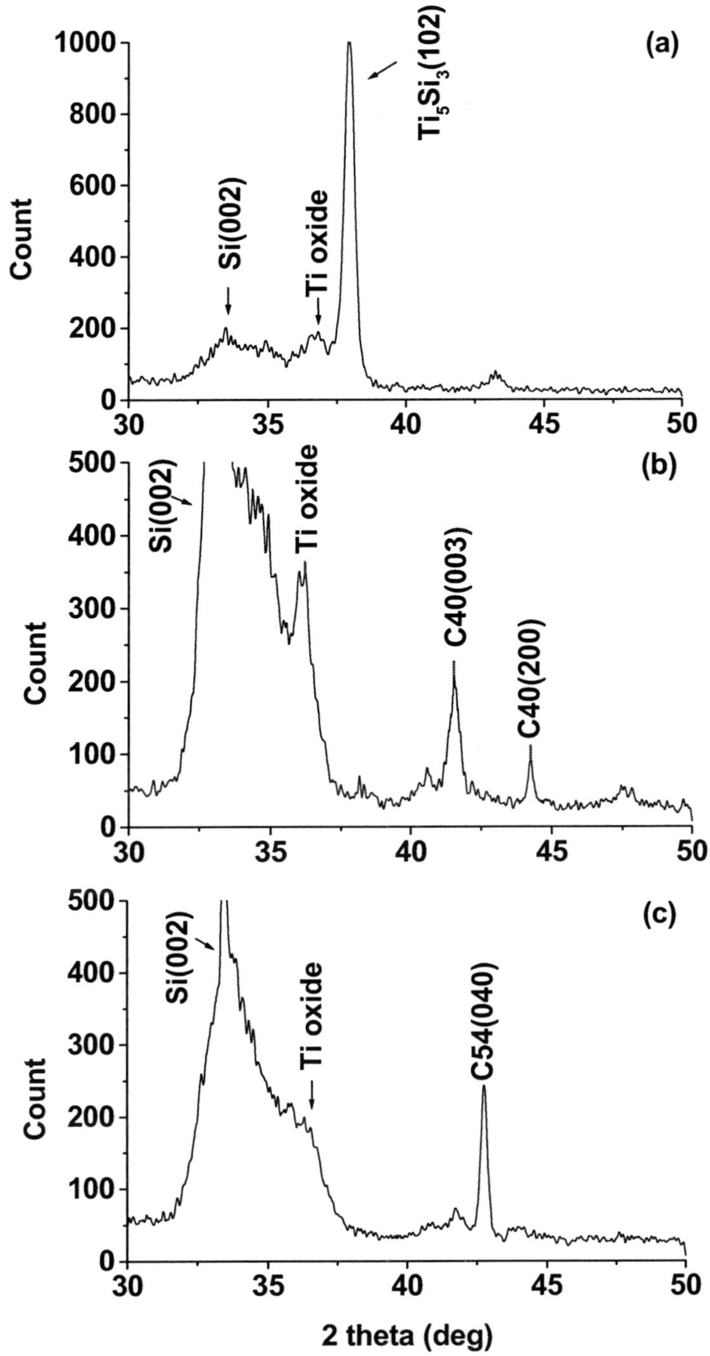

Figure 2. XRD patterns of the Ti/Ta/Si samples annealed at 530℃ (a), 580℃ (b), and 650℃ (c) respectively for 5 min

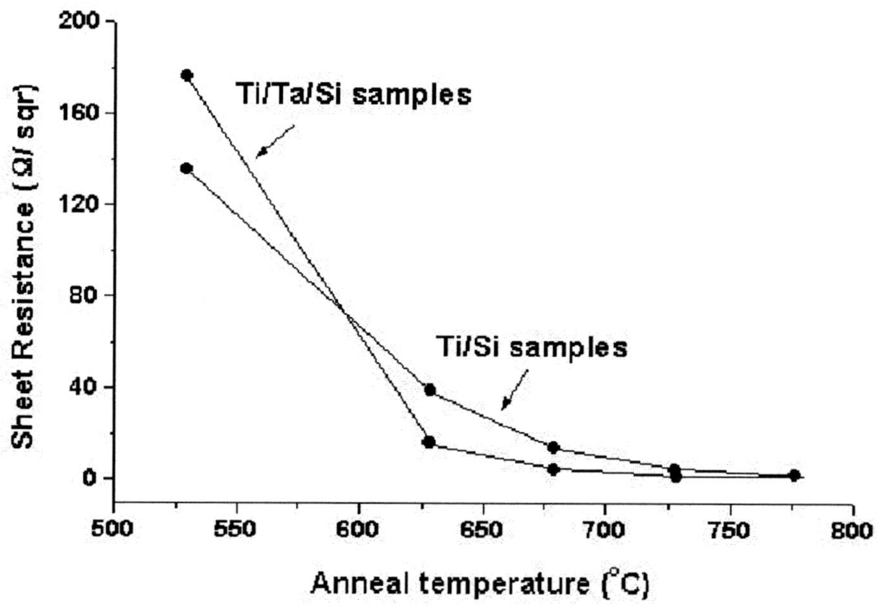

Figure 3. Sheet resistance of the Ti/Si and Ti/Ta/Si samples as a function of annealing temperature

When the Ti/Ta/Si samples are annealed at 580℃, two peaks at 44.6° and 41.8° respectively are found, as shown in Fig.2 (b), and they cannot be originated from C49-$TiSi_2$, C54-$TiSi_2$ or C40-$TaSi_2$ [17]. One explanation for this is that the diffraction peaks are corresponding to the C40-$TiSi_2$ phase, which has the same crystal structure with the C40-$TaSi_2$, and subtle differences in lattice parameters. In this article, we obtain the lattice constants a and c of C40-$TiSi_2$ according to the peak positions of C40 (200) and (003) orientations appearing in our experiment, and compare the results with others' work. All above are listed in Table I. We can see that our experimental results agree well with others, indicating the formation of the single C40 $TiSi_2$ phase when Ti/Ta/Si samples are annealed at high temperatures.

Fig.4 shows the **AES** depth profile of the Ti/Ta/Si sample annealed at 650℃, and it can be seen that in the wide range along the depth of the film, the ratio of Ti to Si is constant, demonstrating the formation of stable titanium silicides. In addition, we find the tantalum atoms mainly aggregate at the $TiSi_2$/Si interface (pointed by the arrow in Fig.4). According to the work of F. La Via *et al* [13, 18], the C40 $TaSi_2$ phase is formed at the interface between $TiSi_2$ and Si substrates when the samples are annealed at high temperature. It is consistent with our **AES** results in Fig. 4. Due to the low lattice mismatch between C40-$TiSi_2$ and $TaSi_2$ (~1.5%), the Ta silicides are of benefit for the formation of the C40-$TiSi_2$, by acting as a "seed" to promote the C40 $TiSi_2$ nucleation. Therefore, the single C40-$TiSi_2$ phase (not found in others' work) is formed easily on the template of the Ta silicides in Ti/Ta/Si samples, while it cannot be found in Ti/Si samples in the absence of the Ta layer.

Table I. Comparison of lattice constants of $TaSi_2$ and $TiSi_2$ in C40 structure between this experiment and others' work (Cu Kα irradiation)

		(003)		(200)	
		2θ (°)	d(Å)	2θ (°)	d(Å)
$TaSi_2$	Exp. (Ref.17)	41.2	2.190	43.7	2.072
$TiSi_2$	Exp. (Ref.13)	41.7	2.164	44.5	2.034
	1/2 (**LDA+GGA**) with α (Ref.13)	41.3	2.186	44.4	2.040
	CBED pattern simulation (Ref.14)	41.5	2.177	44.4	2.039
	GAXRD measurement (Ref.15)	40.9	2.206	44.8	2.022
	Exp. (this work)	41.8	2.161	44.6	2.032

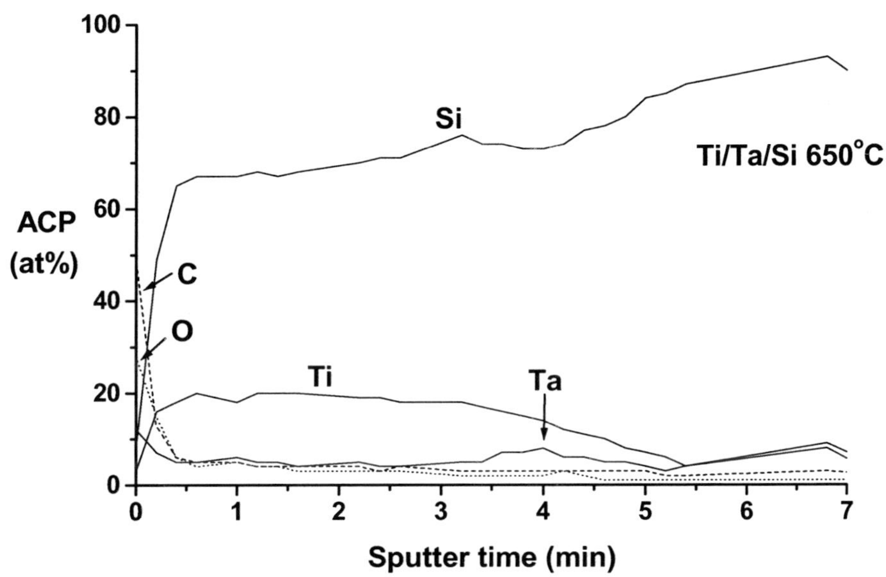

Figure 4. AES depth profile of the Ti/Ta/Si sample annealed at 650℃ for 5 min

The C40-$TiSi_2$ plays a key role in the enhanced formation of C54-$TiSi_2$. In our work, the pure C40-$TiSi_2$ films without any other Ti silicides are formed for the first time, and then the formation temperature of C54-$TiSi_2$ is lowered dramatically by 100℃. From **XRD** data in Fig.2 (b) and (c), it can be found that the C54-$TiSi_2$ with the preferred orientation (040) appears and becomes dominant after the formation of the C40-$TiSi_2$ preferred orientation (003). This demonstrates that the [010] axis of the C54 crystals is collinear with the [001] axis of the C40-$TiSi_2$ crystals when the C54 $TiSi_2$ begins to nucleate on the C40 grains. In fact, the <010> planes of the C54-$TiSi_2$ structure is identical to the basal <001> planes of the C40 $TaSi_2$ [17], and two sets of planes are related by the following relationship: the directions [103] and [100] in the C54 structure are equivalent to the directions $[1\bar{1}0]$ and $[0\bar{1}0]$ in the C40 structure respectively, as shown in Fig.2 of Ref.12. Due to the very small lattice mismatch between C40-$TiSi_2$ and $TaSi_2$, it is reasonable to conclude the <001> planes of C40

$TiSi_2$ are identical to the <010> planes of the C54-$TiSi_2$. According to the lattice parameters of C54-$TiSi_2$ (in Ref.19) and C40-$TiSi_2$ (derived from our experiment results), the lattice mismatch along the C54 [100] or C40 $[0\bar{1}0]$ direction is 2.0%, and the mismatch along the C54 [103] or C40 $[1\bar{1}0]$ direction is 1.8%. The small lattice mismatch between C54 <010> planes and C40 <001> planes will lead to so small an interface energy that the C54 phase is preferred to nucleate and grow in the presence of the C40-$TiSi_2$, blocking the growth of the C49 phase because of the large lattice mismatch between C49 and C40 $TiSi_2$ phase.

The C54-$TiSi_2$, not the C49 phase, directly nucleates and grows "expitaxially" on the pure C40-$TiSi_2$ grains, leading to the dramatic decrease 100℃ in the C54 formation temperature. F. La Via and his co-workers [13] lowered this formation temperature by ~70℃, and they did not obtain the single C40-$TiSi_2$ phase. The differences between the two works are mainly because of the experimental conditions such as the substrates used in two experiments and the methods of thermal processing, as they treated the samples in traditional annealing while the **RTA** was used in this article. It seems that the experimental conditions in this work are of benefit for the formation of the pure C40-$TiSi_2$ films, which account for the larger decrease in the formation temperature of C54-$TiSi_2$. When the high crystal quality C40-$TiSi_2$ films are obtained without any other Ti silicides, the atoms in the crystal are arranged orderly, and the total energy for the whole system will be decreased to some extent. Thus the driving force for the reaction C40-$TiSi_2 \rightarrow$ C54-$TiSi_2$ will be increased, resulting in the lower temperature needed than the system in which the poor crystal quality C40-$TiSi_2$ coexisted with other Ti slicides.

On the other hand, the C49-$TiSi_2$ phase is formed directly after the formation of Ti_5Si_3 due to the absence of the C40-$TiSi_2$ phase for the Ti/Si samples without the Ta interlayer. And the higher annealing temperature is needed, compared with Ti/Ta/Si samples. In the enhancement of the C54-$TiSi_2$ formation, it is very important to limit the nucleation and growth C49 phase and form C54 phase directly, and the C40-$TiSi_2$ phase plays this role well because its <001> planes match the <010> planes of the C54 phase very well and lead to the small interface energy between tow kinds of Ti silicides, which benefits the formation of new C54 phase.

2.2. Epitaxial $CoSi_2$ Films on Si(100) Substrates

In recent years, the research and development of metal silicides have been focusing mainly on minimizing resistance and controlling Si consumption during thin film formation in confined areas. This has lead to a shift from $TiSi_2$ as the most commonly used silicide to $CoSi_2$, which has become a promising candidate for future generation contact materials used in ultra-large scale integrated circuits **(ULSI)**. It is desirable to use epitaxial $CoSi_2$ films in the source/drain areas because they have better uniformity and thermal stability than polycrystalline ones [20]. Although epitaxial $CoSi_2$ films have been grown successfully on Si(111) substates by depositing pure Co films and subsequent annealing at high temperature under ultrahigh vacuum [21], this process failed to produce epitaxial $CoSi_2$ films on Si(100) substrates. Instead, polycrystalline $CoSi_2$ films with (110), (100) and even (221) orientations were fabricated in this way [22, 23]. To settle this problem, a thin Ti interlayer was added as a diffusion barrier, which is called Ti-interlayer mediated epitaxy **(TIME)** [24, 25]. This technique, however, suffers from defects such as lateral encroachment and voids generated under the edge of the oxide [26-28]. Later, oxide mediated epitaxy **(OME)** was developed to overcome the drawbacks of **TIME** [28]. This novel technique is able to grow uniform, epitaxial $CoSi_2$ films on single crystal Si substrates with different orientations and doping levels. However, to maitain the quality of the epitaxial $CoSi_2$ films only 1-3nm thick Co films can be deposited at a time and thicker $CoSi_2$ films have to be obtained by repeated deposition of Co and *in situ* annealing. Recently, a modified **OME** method was proposed where a Ti capping was added on Co/oxide/Si [29, 30]. This process enables the growth of thicker $CoSi_2$ films by *ex situ* annealing, which makes the **OME** method more practical.

Despite all these progress in growing epitaxial $CoSi_2$ films on Si(100) substrates, it is more desirable to achieve this goal simply by direct Co-Si solid phase reaction without using interlayers at all. It should be noticed that in the cases where polycrystalline $CoSi_2$ films were formed by the deposition of pure Co films and subsequent annealing, the substrates were kept at room temperature during the deposition [22, 23, 28, 31]. In this way, the as-deposited Co films are amorphous [32, 33] and the formation of silicides during high temperature rapid thermal annealing **(RTA)** may be the result of the reaction between amorphous Co films and Si substrates. Our previous work has shown that the solid state reaction between crystalline, (002) oriented Co films and Si(111) substrates may be beneficial for the epitaxial growth of $CoSi_2$ on Si(111) substrates [34]. Here, we report the epitaxial growth of $CoSi_2$ films on Si(100) substrates via direct reaction between crystalline, (002) oriented Co films and Si(100) substrates. X-ray diffraction was used to identify the phases and their preferred orientations during the Co-Si reaction. Rutherford backscattering/channeling spectrometry **(RBS)** and high resolution transmission electron microscopy **(HRTEM)** were used to evaluate the quality of the epitaxial $CoSi_2$ films.

Co films with a thickness of 150nm were *e*-gun evaporated onto 3-5 Ω cm p-type Si(100) substrates at a rate of 0.2nm/s. To obtain crystalline Co films, the substrates were kept at 150℃ during the deposition. Then the samples were annealed by **RTA** at various temperatures in a high vacuum of 3×10^{-6}Torr for 2 min.

The **XRD** patterns of the samples annealed at different temperatures are shown in Fig. 5. The as-deposited Co film has a (002) texture, as shown in Fig. 5a. Fig. 5b shows the **XRD** pattern of the sample annealed by **RTA** for 2min at 460℃, which reveals that the Co film exhibits a much stronger (002) preferred orientation but no silicides were formed. After **RTA** at 500℃ for 2min, CoSi with a strong (210) texture was formed (Fig. 5c). From Fig. 5b and c, it can be concluded that CoSi was the first silicide phase to appear and that Co_2Si has been skipped just like the cases of **TIME** and **OME** [28, 35]. These results are in good agreement with Zhang's recent work [36] where the Si substrates were also kept at 150℃ during the deposition. In Fig. 5d, only Si(400) and $CoSi_2$(400) diffraction peaks can be seen, indicating a strong (100) texture of the $CoSi_2$ film after annealing at 850℃. The (220) diffraction peak of $CoSi_2$ disappear completely, in sharp contrast to the strong (220) reflection of the polycrystalline $CoSi_2$ films formed by direct Co-Si reaction reported before [22, 23].

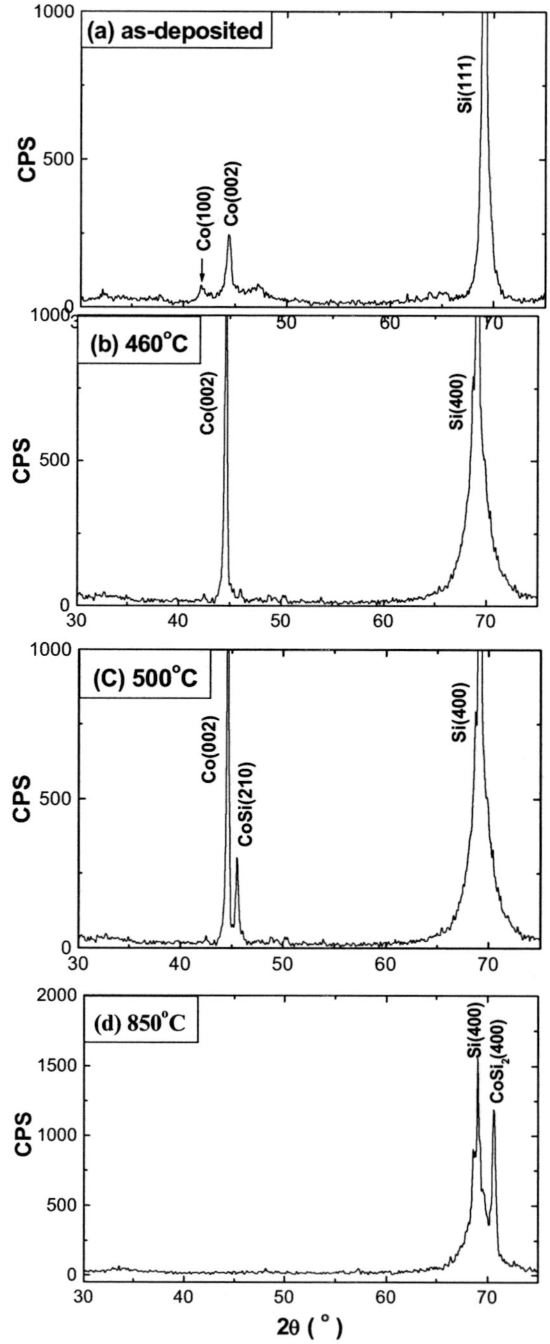

Figure 5. XRD patterns of (a) the as-deposited sample, and the samples annealed by RTA for 2 min at (b) 460℃, (c) 500℃ and (d) 850℃

Fig. 6 shows the channeling and random **RBS** spectra of the sample annealed at 850℃. A 1-mm-diam collimated beam was used in the **RBS** analysis. The minimum channeling yield

χ_{min} was calculated to be 20%. This value is comparable to that of the epitaxial $CoSi_2$ films obtained by **TIME** [37]. The bump pointed by the arrow in Fig. 6 is caused by the overlap of Co and Si signals due to the relatively large thickness of $CoSi_2$ films (~500nm). The **RBS** channeling analysis shows a good epitaxial quality of the $CoSi_2$ films on a macroscopic scale. High resolution electron microscopy was carried out to examine the $CoSi_2$/Si interface and the coherency between the two layers. The structure image of the $CoSi_2$/Si interface shown in Fig. 7 reveals a sharp interface and a complete registry of atoms across the interface, indicating that $CoSi_2$ film is fully coherent with the Si substrate. From the experimental results above we can conclude that good epitaxial $CoSi_2$ films were formed on Si(100) substrates through direct solid phase reaction between the crystalline, (002) oriented Co film and the Si substrate.

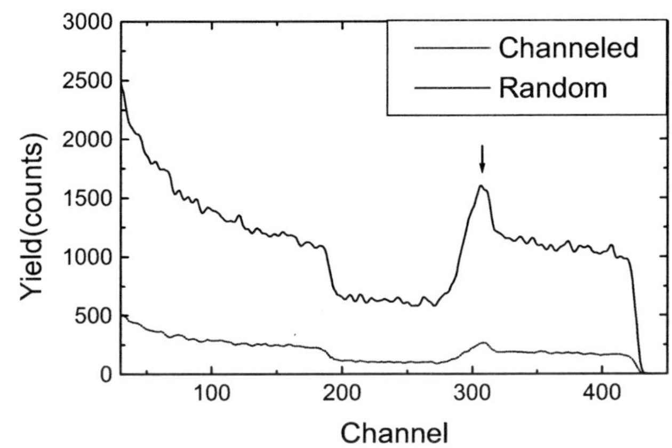

Figure 6. RBS channeled and random spectra of the sample annealed at 850℃

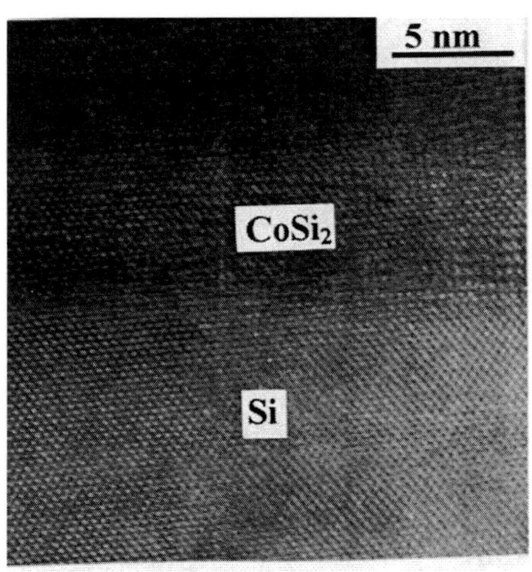

Figure 7. HRTEM image of the $CoSi_2$/Si interface viewed along the Si<110> direction

Why is the solid phase reaction between the textured Co films and Si(100) substrates able to produce epitaxial $CoSi_2$ films while that between amorphous Co films and the Si substrates is not? Here the concept of "thermodynamic barrier" is employed to explain the reason. In **TIME** and **OME**, an interlayer is used as a diffusion barrier which limits the Co-Si reaction until temperatures above 500℃ and allow the precursor phase of Co_2Si to be skipped [20, 30]. The interlayer also retard the nucleation rate of $CoSi_2$ at high temperature annealing [38], which is beneficial for the formation of single crystal $CoSi_2$ films due to the decrease in the number of $CoSi_2$ nuclei. Here we call the interlayer a "kinetic barrier" to the silicidation process. In the case of the reaction between the textured Co film and the Si substrate, the textured Co film itself acts as a "thermodynamic barrier" to the nucleation of silicides compared with its amorphous counterpart. The critical activation energy for the nucleation of silicides is proportional to the quantity $\Delta\sigma^3/\Delta G^2$, where $\Delta\sigma$ stand for the change in interfacial energy [$\Delta\sigma=\sigma$(silicide/Si)+ σ(silicide/Co)-σ(Co/Si)] while ΔG denotes the change in Gibbs free energy [ΔG=G(silicide)-G(Si)-G(Co)] due to the nucleation of silicides [39]. Since the Co film deposited on the hot Si(100) substrate readily exhibits a (002) texture, it is considered that the Co/Si interface energy is reduced to the minimum in this way. Thus σ_t(Co/Si)< σ_a(Co/Si), where footnotes "t" and "a" stand for textured and amorphous Co films, respectively. This leads to $\Delta\sigma_t>\Delta\sigma_a$. On the other hand, it is evident that $G_t(Co)<G_a(Co)$ because amorphous phases have higher free energy than crystalline ones. Noticing that ΔG is a minus value in the silicidation reaction, we have $|\Delta G_t|<|\Delta G_a|$. From the discussion above, we can know that the critical activation energy for the nucleation of silicides in Co(textured)-Si reaction is higher than that in Co(amorphous)-Si reaction, which means the nucleation of silicides is retarded in the Co(textured)-Si reaction. Therefore the textured Co film is equivalent to the interlayer used in **TIME** and **OME** in that both of them hinder the nucleation of silicides. The only difference is that the former do this from a thermodynamic aspect while the latter retard the nucleation kinetically. Indeed, in both cases the intermidiate phase of Co_2Si was skipped, which attest to the equivalent effects of both methods.

2.3. Investigation of NiSi Thin Films

Owing to its line-width-independent, low resistance and relatively low consumption of Si, NiSi has become an attractive candidate for contact material in future generation deep submicron Si devices [40-43]. On the other hand, as the production of transistors reaches gate lengths significantly shorter than 50nm, the formation of cobalt silicide contacts becomes more difficult [44]. So the investigation of nickel silicide has been a hotspot in contact materials field.

2.3.1. Investigation of Thermal Stability of NiSi Films

Due to the lower consumption of Si during the silicidation process, NiSi is more favorable for producing shallow junctions than $CoSi_2$. However, the phase transition from Nisi to high-resistance $NiSi_2$ at temperature greater than 750℃ poses a serious problem [45]. In the past few years, great efforts have been made to study the thermal stability of NiSi films [46-52]. Here, the effect of a thin interlayer of Pt on the thermal stability and texture of NiSi is presented.

An 8nm thin layer of Pt and a 100nm-thick layer of Ni were sequentially e-gun evaporated on the Si(111) substrate, forming a Ni/Pt/Si bilayered film. The base pressure of the vacuum chamber was 1×10^{-8}Torr and during the deposition, the vacuum was better than 3×10^{-8}Torr. The deposition rates were 0.4Å/s for Pt and 2Å/s for Ni. To make a comparison, 100nm-thick Ni films without Pt interlayers were also deposited with the same experimental parameters. Rapid thermal annealing (**RTA**) of the samples was carried out in a high vacuum of 3×10^{-6}Torr at temperatures ranging from 600 to 900℃ for 2 min.

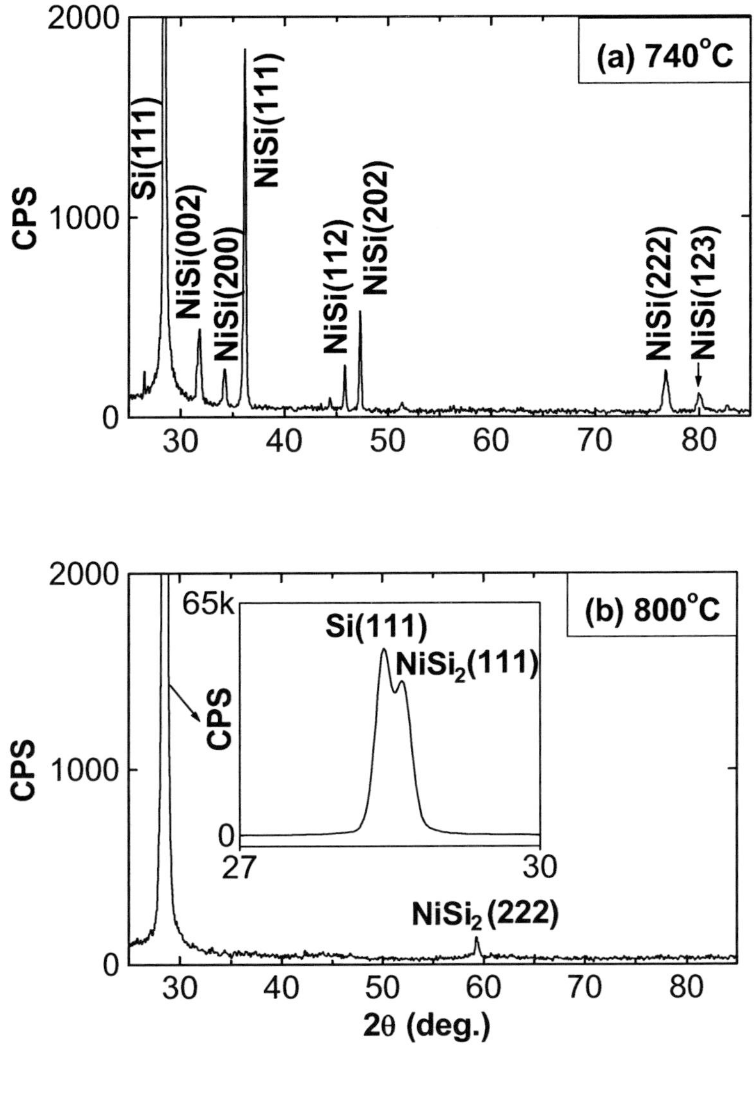

Figure 8. XRD patterns of the Ni/Si samples annealed at (a) 740℃ and (b) 800℃ for 2min

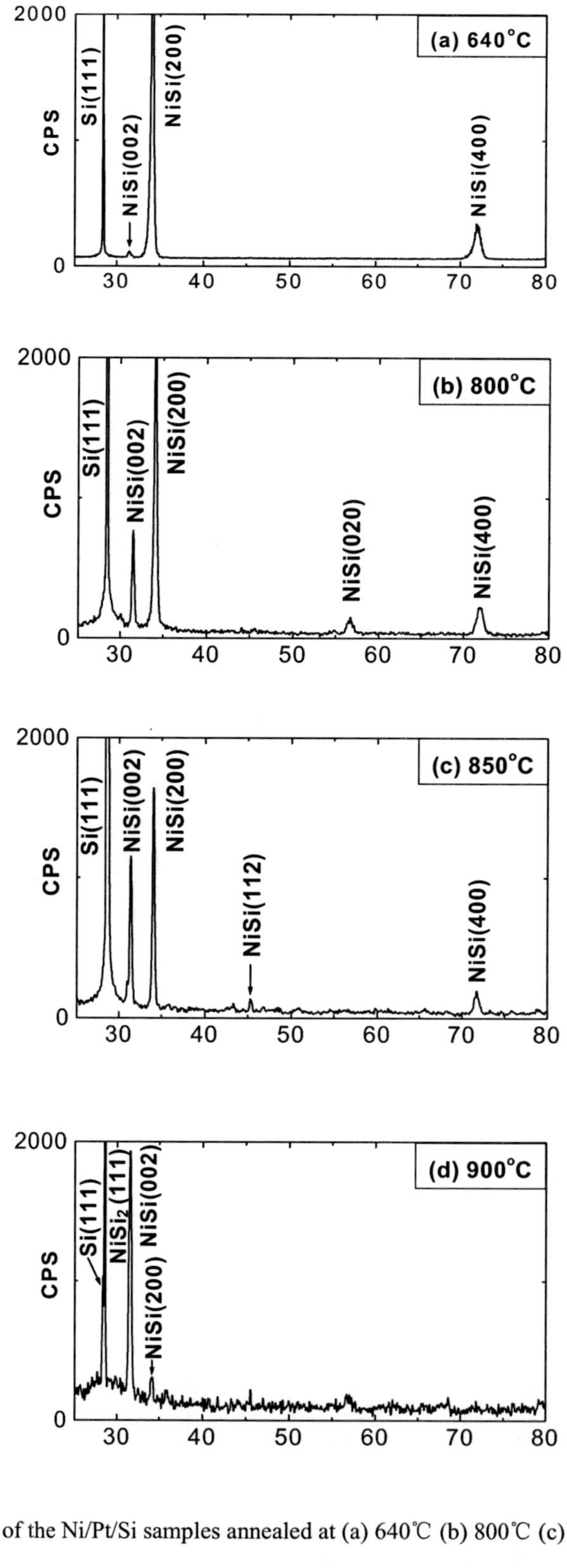

Figure 9. XRD patterns of the Ni/Pt/Si samples annealed at (a) 640℃ (b) 800℃ (c) 850℃ (d) 900℃ for 2min

The **XRD** patterns of Ni/Si samples annealed at different temperatures are shown in Fig. 8. It can be seen that after annealing at 740°C, polycrystalline NiSi films were formed [Fig. 8(a)]. When the annealing temperature rises to 800℃, only peaks corresponding to (111) $NiSi_2$ and (222) $NiSi_2$ planes are present [Fig. 8(b) and its inset], indicating that epitaxial $NiSi_2$ films were formed. From these data, it is concluded that the transition from NiSi to $NiSi_2$ is completed at 800°C. For Ni/Pt/Si samples, on the other hand, this phase transition was not completed even after annealing at 900°C [Fig. 9(d)], indicating a remarkable enhancement of thermal stability of NiSi. This is also confirmed by the sheet resistance measurement [Fig. 10]. For Ni/Si samples, a dramatic increase in sheet resistance from 0.60 Ohm/Square to 1.40 Ohm/Square took place after annealing at 800 C due to the formation of NiSi2 as shown in the corresponding XRD pattern [Fig. 8 (b)]. For Ni/Pt/Si samples, this dramatic increase in sheet resistance did not take place until they were annealed at 900 C. Even after annealing at this temperature, the sheet resistance of the sample (1.0 Ohm/Square) was still lower than that of the pure NiSi2 film because the NiSi-to-NiSi2 transition was not completed. Thus it is concluded that the addition of a thin Pt interlayer raises the NiSi-to-$NiSi_2$ transition temperature by over 100°C.

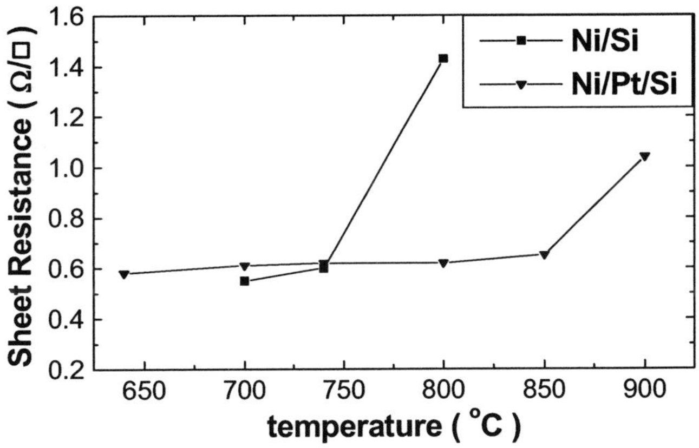

Figure 10. Sheet resistance of the Ni/Si and Ni/Pt/Si samples as a function of the annealing temperature

Fig. 11 shows the TEM analysis of the sample annealed at 640°C. From the **SAED** pattern at the NiSi/Si interface [Fig. 11(c)] it was determined that the orientation relationship between the NiSi film and the Si substrate is NiSi(100)||Si(111) and NiSi[$0\bar{1}0$]||Si[$01\bar{1}$], in agreement with the **XRD** data shown in Fig. 9(a). The **HRTEM** image in Fig. 11(b) reveals a well-defined interface with atoms orderly arranged on both sides. These results suggest that the NiSi films with a texture of NiSi(100)||Si(111) may have much lower interfacial energy than polycrystalline NiSi films. The transformation from NiSi to $NiSi_2$ is a nucleation-controlled reaction [52-54]. The activation energy for nucleation $\Delta G^* \cong \Delta\sigma^3/\Delta G^2$, where $\Delta\sigma$ and ΔG stand for the change in the interfacial energy and Gibbs free energy during the reaction of $NiSi+Si \rightarrow NiSi_2$, respectively [52,53]. The reduction of NiSi/Si interfacial energy due to the texture of NiSi(100)||Si(111) leads to an increase in $\Delta\sigma$, which can cause a significant increase in ΔG^* as this quantity is proportional to $\Delta\sigma^3$. Thus the nucleation of

$NiSi_2$ is effectively retarded and the thermal stability of NiSi films is markedly enhanced. It is interesting to see that with the increase of temperature, the preferred orientation of NiSi films changes from NiSi(100)||Si(111) to NiSi(001)||Si(111) [Fig. 9 (a)-(d)]. While the texture of NiSi(100) was also observed in Mangelinck's recent work [52] where Pt was added into NiSi films by using a Ni(Pt) alloy target, such conversion in the texture of NiSi films at an elevated temperature was not reported. From Fig. 9 (a)-(d), it seems that the nucleation of $NiSi_2$ begins only after this conversion in NiSi texture. It is probable that in our case the activation energy ΔG^* is so large for the phase transformation from (100) oriented NiSi films to $NiSi_2$ films that there has to be a transition in NiSi texture from NiSi(100)||Si(111) to NiSi(001)||Si(111) before the nucleation of $NiSi_2$ in order to increase σ(NiSi/Si) and decrease ΔG^*.

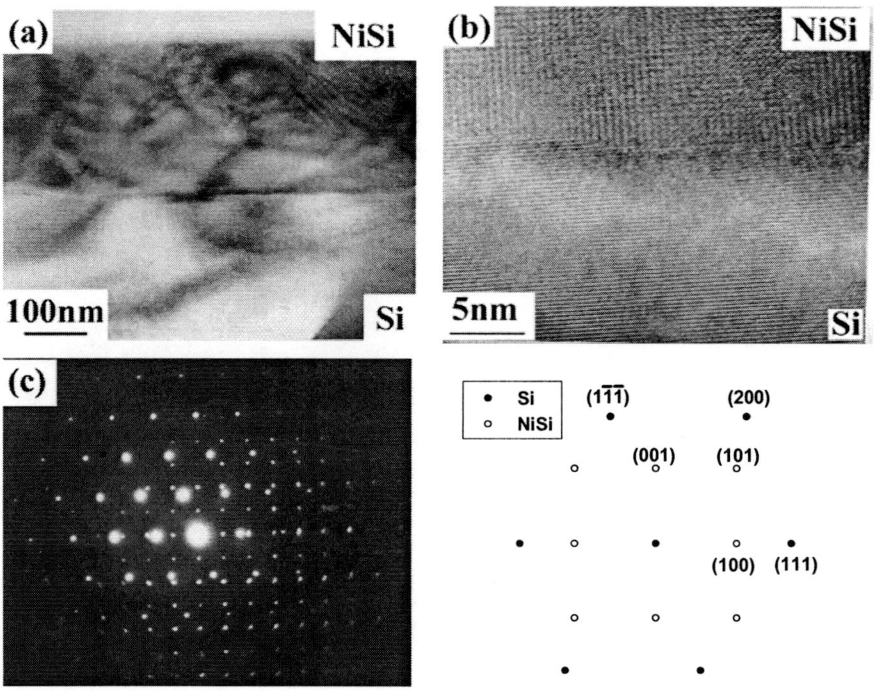

Figure 11. TEM analysis of the Ni/Pt/Si sample annealed at 640℃: (a) Cross section TEM morphology, (b) HRTEM image at the interface, and (c) SAED pattern at the NiSi/Si interface

As shown in the **XRD** patterns of Ni/Pt/Si samples (Fig. 9), no peak corresponding to PtSi appears. By comparing the positions of NiSi diffraction peaks in Fig. 8 with those in Fig. 9, one can notice subtle differences in peak positions. These are listed in Table II. The reason for this discrepancy is attributed to the formation of NiSi-PtSi solid solution. Since the original thickness of Pt and Ni films is 8nm and 100nm, respectively, it can be calculated that the atomic percentage of PtSi in this solid solution is 5.5%. Both NiSi and PtSi have an orthorhombic MnP structure, with lattice constants a=5.233Å, b=3.258Å, c= 5.659Å for NiSi [55] and a=5.595Å, b=3.603Å, c=5.932Å for PtSi [56]. By applying a Vegard's law, the lattice parameters of $Ni_{0.945}Pt_{0.055}Si$ solid solution should be a=5.253Å, b=3.277 Å and

c=5.674Å. This result agrees quite well with the positions of NiSi(002), (200) and (123) diffraction peaks in annealed Ni/Pt/Si samples as listed in Table II. Moreover, it is shown in **AES** depth pofile (Fig. 12) that although Pt was deposited before Ni, it distribute evenly through the depth of the Ni(Pt)Si film after annealing at 640°C. From the experimental data above, we can conclude that PtSi solute in NiSi solvent in such a manner that Pt atoms substitute some of the Ni atoms in NiSi lattice. The MnP-typed structure of NiSi can be regarded as a distortion of the NiAs structure (hexagonal, space group $P6_3/mmc$) [52, 54]. If the ratio of c/b equals $\sqrt{3}$ in MnP-typed structure, then the structure can also be expressed in terms of space group $P6_3/mmc$ when the unit-cell dimensions (a, b, c) are linked to those of space group $P6_3/mmc$ (a_h, c_h) by the following relationships: $a_h = b$, $c_h = a$ [52, 54]. Thus the (100) planes of MnP-typed structure become (0001) planes of hexagonal NiAs structure. In our case, the addition of 5.5 at. % of Pt adjusts the ratio of c/b from 1.7370 of pure NiSi to 1.7315($\approx\sqrt{3}$) of $Ni_{0.945}Pt_{0.055}Si$. Then the texture of Ni(Pt)Si(100)||Si(111) can also be considered as hexagonal structure (0001)||Si(111). As both (0001) planes in hexagonal structure and (111) planes in Si are close packed planes, such a texture is beneficial for the reduction of interfacial energy. With the discussion above, we get a possible reason for the texture of NiSi(100)||Si(111) in annealed Ni/Pt/Si samples.

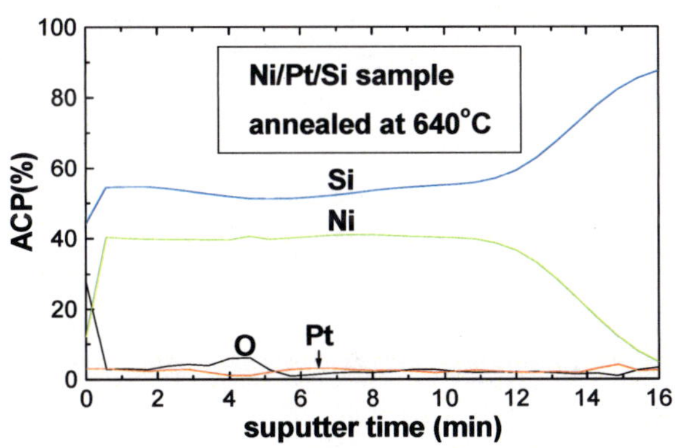

Figure 12. AES depth profile of the Ni/Pt/Si sample annealed at 640℃ for 2 min

Table II. Comparison of XRD peak positions of pure NiSi and Ni(Pt)Si solid solution. (Cu Kα irradiation)

	(002)	(200)	(123)
	2θ (°) d(Å)	2θ (°) d(Å)	2θ (°) d(Å)
NiSi [a)]	31.62 2.829	34.26 2.617	79.94 1.200
Ni(Pt)Si [a)]	31.52 2.838	34.14 2.626	79.34 1.208
Ni(Pt)Si [b)]	31.53 2.837	34.14 2.627	79.50 1.206

a) experimental data;
b) calculated data using Vegard's law

In conclusion, the thermal stability of NiSi films was significantly improved by using a thin Pt interlayer. Detailed study on the **XRD** data shows PtSi and NiSi form a solid solution following a Vegard's law, which adjust the lattice constant ratio c/b to $\sqrt{3}$ and may account for the texture of NiSi(100)||Si(111). It was also found in Ni/Pt/Si samples that a transition in NiSi texture from NiSi(100)||Si(111) to NiSi(001)||Si(111) took place before the nucleation of $NiSi_2$. The enhanced stability of NiSi films and the transition in NiSi texture is attributed to the reduction in the interfacial energy due to the formation of the (100) textured NiSi film and subsequent augment in activation energy for the nucleation of $NiSi_2$.

2.3.2. Effect of Film Thickness on NiSi-to-NiSi₂ Transition Temperature

As the dimensions of devices scale down, the thickness of silicide layers decreases, too. According to the International Technological Roadmap for Semiconductors (ITRS) of 2003, the proposed silicide thickness is 10nm for the line-width of 45nm [59]. Therefore it is of much interest to study the variation in the thermal stability of NiSi films as the film thickness goes down. Here, the effect of film thickness on the NiSi-to-$NiSi_2$ transition temperature in Ni/Pt/Si(100) system is investigated.

To study the film thickness dependence of the NiSi-to-$NiSi_2$ transition temperature in Ni/Pt/Si(100) system, three sets of samples with the same Ni:Pt ratio but different film thickness were prepared by e-gun evaporation: (A): Ni(50nm)/Pt(4nm)/Si(100), (B): Ni(25nm)/Pt(2nm)/Si(100) and (C): Ni(12.5nm)/Pt(1nm)/Si(100). The base pressure of the vacuum chamber was 1×10^{-8}Torr. Then the samples were subjected to rapid thermal annealing (RTA) for 3 min. in a high vacuum of 3×10^{-6}Torr at temperatures ranging from 750 to 900℃ to test the thermal stability of NiSi films in these samples. X-ray diffraction (XRD) and sheet resistance measurement were employed to determine the NiSi-to-$NiSi_2$ transition temperatures in the samples. Auger electron spectra (AES) was used to investigate the composition of the films

Fig. 13(a) shows the XRD patterns of sample A [Ni(50nm)/Pt(4nm)/Si(100)] annealed at different temperatures. No $NiSi_2$ was formed after RTA at 800℃. However, the nucleation of $NiSi_2$ began at an elevated temperature of 840℃. For sample B [Ni(25nm)/Pt(2nm)/Si(100)], the NiSi-to-$NiSi_2$ transition did not begin until after annealing at 880℃, as shown in Fig. 13(b). Fig. 13(c) shows the XRD patterns of sample C [Ni(12.5nm)/Pt(1nm)/Si(100)] after RTA. No diffraction peaks of $NiSi_2$ appeared even after annealing at 880℃. As revealed in our previous work[58], in Ni/Pt/Si bi-layered system NiSi and PtSi form a solid solution $Ni_{1-x}Pt_xSi$ following Vegard's law after annealing. This would lead to noticeable shifts in the XRD peak positions of NiSi. Thus, the peak positions of Ni(Pt)Si films can reflect the atomic percentage of Pt in NiSi lattice. Table III shows the peak positions of NiSi(211) of samples A-C annealed at 800℃. The peak positions are determined using a parabolic top method. The basically same peak positions in Table III reveal that the contents of Pt in samples A-C are identical as expected. This is also confirmed by AES analysis of these three sets of samples. Therefore the varied NiSi-to-$NiSi_2$ transition temperatures in samples A-C is a result of different film thickness instead of differed Pt percentage in the samples.

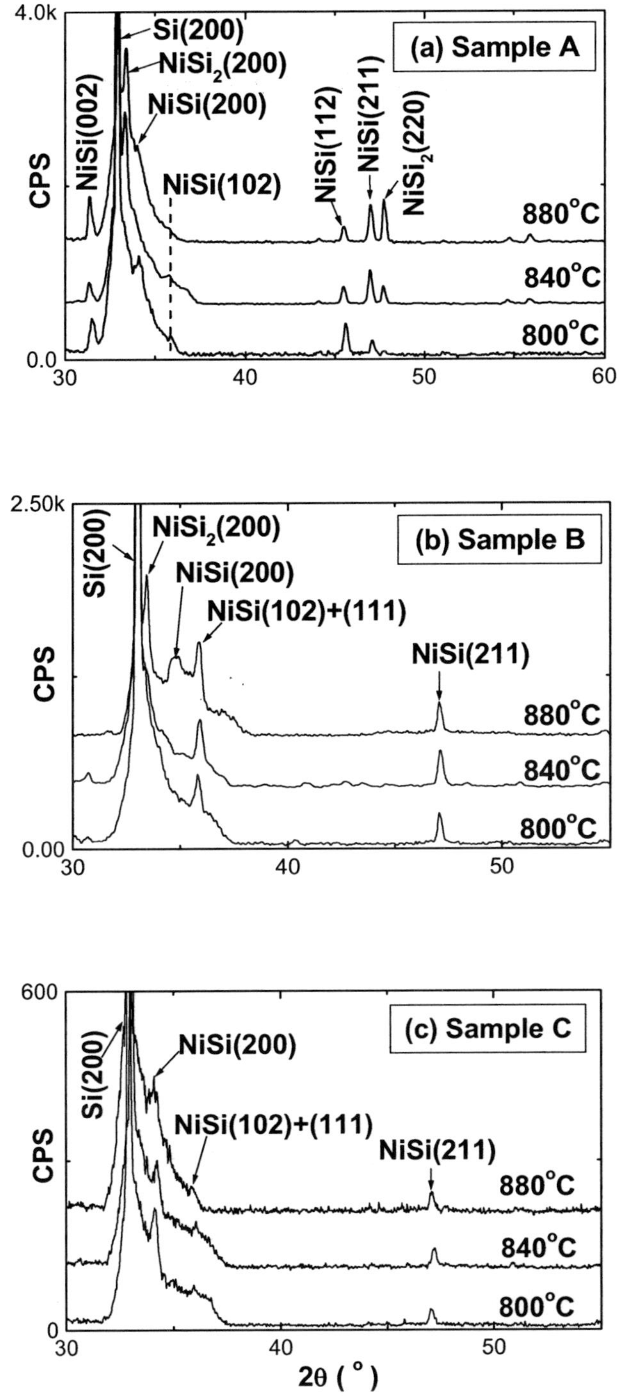

Figure 13. XRD patterns (Cu $K\alpha$ source, λ=1.5415Å) of samples annealed between 800 and 900℃ for 3 min: (a) sample A Ni (50nm)/Pt(4nm)/Si(100), (b) sample B Ni(25nm)/Pt(2nm)/Si(100), and sample C Ni(12.5nm)/Pt(1nm)/Si(100)

Table III. XRD peak positions of NiSi(211) in samples A-C after RTA at 800℃. The peak positions are determined using a parabolic top method

Samples	A	B	C
2θ of NiSi(211)	47.142°	47.148°	47.145°

The film thickness effect on the thermal stability of NiSi formed in Ni/Pt/Si(100) system is also confirmed by the sheet resistance measurement. Fig. 14 shows the sheet resistance of samples A-C in relation to the annealing temperatures. As the resistivity of NiSi films is almost the same while the film thickness ratio is 4:2:1 in samples A-C, the sheet resistance of the NiSi films in these three sets of samples approximately follows the ratio of 1:2:4. After annealing at 840℃, the sheet resistance of sample A increases from 1.7 Ohm/Square to 2.3 Ohm/Square. Referring to the XRD data in Fig. 13(a), it is concluded that the increase in sheet resistance results from the formation of $NiSi_2$. For sample B, such an augment in sheet resistance occurs at a higher annealing temperature of 880℃. In the case of sample C, the sheet resistance remains at 6.8 Ohm/Square even after annealing at 880℃. With the above experimental results, we can conclude that the NiSi-to-$NiSi_2$ transition temperature increases with the decrease of the film thickness in Ni/Pt/Si(100) system. This property indicates a good potential for the application of Ni(Pt)Si films in utra-shallow junctions of ULSI.

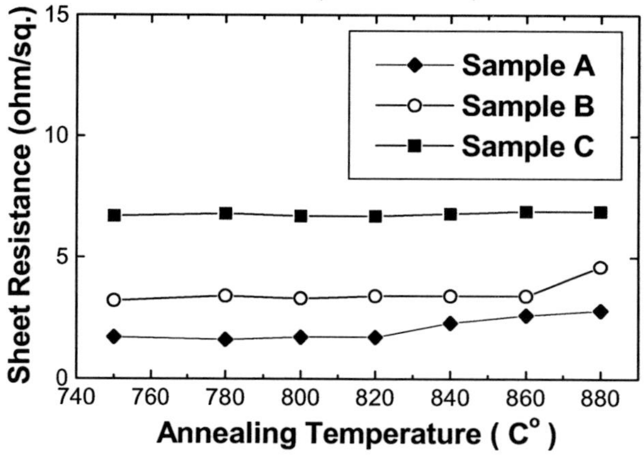

Figure 14. Sheet resistance of samples A-C in relation to the annealing temperatures

Such film thickness dependence of silicide formation has also been observed in Ti-Si and Co-$Si_{1-x}Ge_x$ solid phase reactions in previous literatures. Jeon *et al* [60] reported an increase in the C49-to-C54 $TiSi_2$ transition temperature as the initial Ti film thickness goes down from 40nm to 10nm. Boyanov *et al* [61] reported incomplete CoSi-to-$CoSi_2$ conversion during the solid phase reaction between Co and $Si_{0.79}Ge_{0.21}$ when the initial Co film thickness is below 35nm. In the present work, we've found that the NiSi-to-$NiSi_2$ transition temperature increases with the decrease of the film thichness in Ni/Pt/Si(100) system. The transitions of C49-to-C54 $TiSi_2$, CoSi-to-$CoSi_2$ and NiSi-to-$NiSi_2$ have one thing in common: they are all nucleation controlled reactions characterized by a small Gibbs free energy change ΔG of the reactions and a large activation energy for nucleation of the new phases [62-64]. This would also lead to a relatively large critical nucleation radius r^* which is proportional to 1/|ΔG|. Only the nuclei larger than r^* are able to grow spontaneously. Therefore when the film thickness decreases to the same order of magnitude as r^*, the nucleation of these disilicide phases would be retarded by the limitation of the nucleus size in the direction perpendicular to the film. Next we'll have a more detailed discussion on the NiSi-to-$NiSi_2$ transition in Ni/Pt/Si(100) system.

The NiSi-to-$NiSi_2$ transition is a typical nucleation controlled reaction [64]. Since the range of annealing temperature is mainly between 800 and 900℃ in our experiments, we'll use the thermodynamic data at 1100K in the following discussions. Even at a high temperature of 1100K, the Gibbs free energy change ΔG_0 for the reaction

$$NiSi + Si \rightarrow NiSi_2 \tag{1}$$

is only -2.35kcal/mol or -9.83kJ/mol [64]. From the lattice structure (cubic, CaF_2 structure) and the lattice constant (*a*=5.406 Å) of $NiSi_2$ [65], we can calculate that the Gibbs free energy change per unit volume of $NiSi_2$ in reaction (1) is ΔG_{0V}= - 413J/cm^3. The critical nucleation radius r^* is given as [64]

$$r^* = \frac{2b\Delta\sigma_s}{3a\,|\,\Delta G_V\,|} \tag{2}$$

where $\Delta\sigma_s$ and ΔG_V respectively denote the change in interfacial energy per unit area of the interface and the change in Gibbs free energy per unit volume of the new phases for the reaction, while *a* and *b* are constants that depend on the shape of the nucleus. Taking $\Delta\sigma_s$ to be of the order of 10^{-4}/cm^2 [64], we estimate that the critical nucleation radius r^* for $NiSi_2$ is of the order of several nanometers in reaction (1). With the formation of Ni(Pt)Si solid solution in our case, however, the reaction to form $NiSi_2$ changes into[3]

$$Ni_{1-x}Pt_xSi + (1-x)Si \rightarrow (1-x)NiSi_2 + xPtSi$$

Compared with equation (1), a new phase PtSi accompanies the formation of $NiSi_2$ in equation (3). This would lead to new phase boundaries and increase the $\Delta\sigma_s$ of the reaction. On the other hand, the formation of $Ni_{1-x}Pt_xSi$ solid solution is expected to induce a relatively large decrease in the Gibbs free energy of NiSi since NiSi and PtSi can form a solid solution in the full range of concentration [57, 66]. As a very rough estimate, we assume an ideal mixture of NiSi and PtSi in Ni(Pt)Si solid solution and only consider the effect of mixing entropy ΔS_{mix} on the Gibbs free energy change of reaction (3) compared with reaction (1). Under this assumption, the free energy of $Ni_{1-x}Pt_xSi$ is given as follows:

$$\Delta G^f(Ni_{1-x}Pt_xSi) = (1-x)G^f(NiSi) + xG^f(PtSi) - T\Delta S_{mix}$$

where mixing entropy ΔS_{mix} is given by

$$\Delta S_{mix} = -R[x \ln x + (1-x)\ln(1-x)]$$

Therefore we can calculate that the Gibbs free energy change of equation (3) is

$$\begin{aligned}\Delta G_1 &= (1-x)[G^f(NiSi_2) - G^f(NiSi) - G^f(Si)] + T\Delta S_{mix} \\ &= (1-x)\Delta G_0 + T\Delta S_{mix}\end{aligned}$$

In our case, it can be calculated from the initial thickness of Ni and Pt films that the atomic percentage of Pt in $Ni_{1-x}Pt_xSi$ solid solution is x=0.055 [58]. From equation (5), the term $T\Delta S_{mix}$ is determined to be 2.0kJ/mol at 1100K, which is of the same order of magnitude as ΔG_0. This alone would give rise to a significant reduction in the driving force of NiSi-to-$NiSi_2$ transition. As the above estimate neglects the formation of bonding in $Ni_{1-x}Pt_xSi$ which can lead to a further reduction in the free energy of the solid solution, the real decrease in $|\Delta G|$ of NiSi-to-$NiSi_2$ transition due to the presence of $Ni_{1-x}Pt_xSi$ would be even larger. Therefore the existence of $Ni_{1-x}Pt_xSi$ solid solution increases the interfacial energy change and decreases the absolute value of the Gibbs free energy change for the NiSi-to-$NiSi_2$ transition. According to equation (2), this would lead to a significant increase in r^* compared with reaction (1). Although we are unable to calculate the value of r^* due to the lack of adequate interfacial energy data and an exact calculation of the Gibbs free energy change of equation (3), we estimate that the critical nucleation radius r^* of $NiSi_2$ may increase from the order of several nanometers in reaction (1) to ~10nm in reaction (3) due to the presence of $Ni_{1-x}Pt_xSi$ solid solution. Therefore when the initial Ni film thickness decreases from 50nm to 12.5nm, the nucleation of $NiSi_2$ is more and more limited by the thickness of the films and the NiSi-to-$NiSi_2$ transition temperature subsequently increases.

3. Conclusion

Much progress has been made in researching solid-state reaction of metal-silicon. In this article, we discussed three generations contact materials in CMOS technology. For the first generation contact material, i.e. C54-$TiSi_2$, we compared the Ti/Ta/Si bilayered system and the Ti/Si system in the formation of C54-$TiSi_2$, and observed the single intermediate C40-$TiSi_2$ phase after RTA in the Ti/Ta/Si samples. Due to the small lattice mismatch (<0.3%) in the matched planes of C40 and C54 $TiSi_2$, the C54-$TiSi_2$ was formed directly by skipping the C49 phase when the Ti/Ta/Si samples were annealed further. Then the formation temperature of C54-$TiSi_2$ was lowered by nearly 100℃ compared with Ti/Si samples without the Ta layer. For the second generation contact material, we investigated the epitaxial $CoSi_2$ film which was formed on Si(100) substrate through direct reaction between the textured Co film and the Si substrate. Both RBS channeling analysis and HRTEM reveal good epitaxial quality of the $CoSi_2$ films. The reason for the improvement in the epitaxy is explained in terms of "thermodynamic barrier" caused by the textured Co film. For the third generation contact material, we discussed the thermal stability of NiSi films in details. The enhanced stability of NiSi films and the transition in NiSi texture is attributed to the reduction in the interfacial energy due to the formation of the (100) textured NiSi film and subsequent augment in activation energy for the nucleation of $NiSi_2$. On the other hand, the film thickness dependence of the NiSi-to-$NiSi_2$ transition temperature in Ni/Pt/Si(100) system was discussed in details. We've found that the thermal stability of NiSi films improves with the decrease of film thickness in Ni/Pt/Si(100) system. The experimental results are qualitatively explained in terms of classical nucleation theory. The enhancement of the thermal stability of Ni(Pt)Si films with the decrease of film thickness reveals a good potential for the application of Ni(Pt)Si in ultra-shallow junctions of ULSI where very thin silicide films are required.

ACKNOWLEDGEMENTS

This work was supported by the National Natural Science Foundation of China. The authors thank Mr. R.R. He for performing the HRTEM analysis. The authors also thank Dr. P. S. Lee of National University of Singapore for useful discussions.

REFERENCES

[1] J. P. Gambino; E. G. Colgan *Mater. Chem. Phys*. 1998, 52, 99-146.
[2] Steegen; K. *maex Mater. Sci. Eng. R* 2002, 38, 1-53.
[3] M. –A. Nicolet, S. S. Lau, in: N. G. Einspruch, G. B. Larrabee (Eds.), *VLSI Electronics Microstructure Science,* Academic Press, New York, 1983.
[4] R. Beyers; R. Sinclair *J. Appl. Phys*. 1985, 57, 5240-5245.
[5] F. M. D'Heurle, P. Gas, I. Engstrom, S. Nygren, M. Ostling and C. S. Peterson, *IBM Report RC 11151 No. 50067* [New York: Yorktown Heights], 1985.
[6] J. A. Kittl; D. A. Prinslow; P. P. Apte; M. F. Pas *Appl. Phys. Lett*. 1995, 67, 2308-2310.
[7] S. Privitera; F. La Via; M. G. Grimaldi; E. Rimini *Appl. Phys. Lett*. 1998, 73, 3863-3865.
[8] R. W. Mann; G. L. Miles; T. A. Knotts; D. W. Rakowshi; L. A. Clevenger; J. M. E. Harper; F. M. D'Heurle; C. Cabral C *Appl. Phys. Lett*. 1995, 67, 3729-3731.
[9] C. Cabral; L. A. Clevenger; J. M. E. Harper; F. M. D'Heurle; R. Roy; C. Lavoie; K. Saenger *Appl. Phys. Lett*. 1997, 71, 3531-3533.
[10] Mouroux; S. L. Zhang; C. S. Peterson *Phys. Rev. B* 1997, 56, 10614-10620.
[11] J. A. Kittl; M. A. Gribelyuk; S. B. Samavedam *Appl. Phys. Lett*. 1998, 73, 900-902.
[12] Mouroux; T. Epicier; S. L. Zhang; P. Pinard *Phys. Rev. B* 1999, 60, 9165-9168.
[13] F. La Via; F. Mammoliti; G. Corallo; M. G. Grimaldi; D. B. Migas; L. Miglio *Appl. Phys. Lett*. 2001, 78, 1864-1866.
[14] K. Li; S. Y. Chen; Z. X. Shen *Appl. Phys. Lett*. 2001, 78, 3989-3991.
[15] T. Yu; S. C. Tan; Z. X. Shen; L. W. Chen; J. Y. Lin *Appl. Phys. Lett*. 2002, 80, 2266-2268.
[16] R. N. Wang; J. Y. Feng; Y. Huang *J. Crys. Grow*. 2003, 253, 280-285.
[17] M. E. Schlesinger *J. Phase Equilibria* 1994, 15, 90-94.

[18] F. La Via; S. Privitera; F. Mammoliti; M. G. Grimaldi *Microelectronic Engineering* 2002, 60, 197-203.
[19] JCPDS, 35-785, (1998).
[20] R. T. Tung; F. Schery *Appl. Phys. Lett.* 1995, 67, 2164-2166.
[21] R. T. Tung *Mater. Chem. Phys.* 1992, 32, 107-111.
[22] J. R. Jimenez; L. J. Schowalter; L. M. Hsiung; K. Rajan; S. Hahimoto; R. D. Thomson; S. S. Lyer *J. Vac. Sci. Technol. A* 1990, 8, 3014-3018.
[23] S. M. Yalisove; R. T. Tung; D. Loretto *J. Vac. Sci. Technol. A* 1989, 7, 1472-1474.
[24] M. Lawrence A. Dass; David B. Fraser; Chih-shih Wei *Appl. Phys. Lett.* 1991, 58, 1308-1310.
[25] T. I. Selinder; D. J. Miller; K. E. Gray *Appl. Phys. Lett* 1995, 67, 1597-1599.
[26] R. T. Tung; F. Scherey *Appl. Phys. Lett* 1995, 67, 2164-2166.
[27] J. S. Byun; J. M. Seon; K. S. Youn; H. Hwang; J. W. Park; J. J. Kim *J. Electrochem. Soc.* 1996, 143, L56-L58.
[28] R. T. Tung; *Appl. Phys. Lett.* 1996, 68, 3461-3463.
[29] G. B. Kim; J. S. Kwak; H. K. Baik *J. Vac. Sci. Technol. B* 1999, 17, 162-165.
[30] G. B. Kim; J. S. Kwak; H. K. Baik *J. Appl. Phys.* 1999, 85, 1503-1507.
[31] W. D. Chen; Y. D. Cui; C. C. Hsu *J. Appl. Phys.* 1991, 69, 7612-7619.
[32] J. Pelleg; S. Zalkind; L. Zevin; B. M. Ditchek *Thin Solid Films* 1994, 249, 126-131.
[33] S. Zalkind; Joshua Pelleg; L. Zevin; B. M. Ditchek *Thin Solid Films* 1994, 249, 187 - 194.
[34] J. F. Liu; J. Y. Feng; B. Li; J. Zhu *J. Crys. Growth.* 2000, 209, 795-800.
[35] F. Hong; G. A. Rozgonyi *Appl. Phys. Lett.* 1994, 64, 2241-2243.
[36] H. Zhang; J. Poole; R. Eller; M. Keefe *J. Vac. Sci. Technol. A* 1999, 17, 1904-1910.
[37] G. B. Kim; J. S. Kwak; H. Baik *J. Appl. Phys.* 1997, 82, 2323-2328.
[38] M. Falke; B. Gebhardt; G. Beddies; S. Teichert; H.-J. Hinneberg *Thin Solid Films* 1998, 336, 201-204.
[39] D. Mangelinck; J. Y. Dai; J. S. Pan; S. K. Lahiri *Appl. Phys. Lett.* 1999, 75, 1736-1738.
[40] J. P. Gambino; E. G. Colgan *Mater. Chem. Phys.* 1998, 52, 99-146.
[41] F. Deng; R. A. Johnson; P. M. Asbeck; S. S. Lau; W. B. Dubbelday; T. Hsiao; J. Woo *J. Appl. Phys.*1997, 81, 8047-8051.
[42] J. Chen; J. P. Colinge; D. Flandre; R. Gillon; J. P. Raskin; D. Vanhoenacker *J. Electrochem. Soc.* 1997, 144, 2437-2442.
[43] T. Ohguro; S. Nakajima; M. Koike; T. Morimoto; A. Nishiyama; Y. Ushiku; T. Yoshitomi; M. Saito; H. Iwai *IEEE Trans. Electron Devices* 1994, 41, 2305-2317.
[44] C. Lavoie; F. M. d'Heurle; C. Detavernier; C. Cabral Jr. *Microelectronic Eng.* 2003, 70, 144-157.
[45] J. F. Liu; H. B. Chen; J. Y. Feng; J. Zhu *Appl. Phys. Lett.* 2000, 77, 1-3.
[46] D. Mangelinck; P. Gas; A. Grob; B. Pichard; O. Thomas *J. Appl. Phys.* 1996, 79, 4078-4086.
[47] M. C. Poon; C. H. Ho; F. Deng; S. S. Lau; H. Wong *Microelectron. Reliab.* 1998, 38, 1495-1498.
[48] M. C. Poon; M. Chan; W. Q. Zhang; F. Deng; S. S. Lau *Microeletron. Reliab.* 1998, 38, 1499-1502.
[49] L. W. Cheng; S. L. Chang; J.Y. Chen; L. J. Chen; B. Y. Tsui *Thin Solid Films* 1999, 356, 412-416.

[50] C. J. Tsai; P. L. Chung; K. H. Yu *Thin Solid Films* 2000, 365, 72-76
[51] M. C. Poon; F. Deng; M.Chan; W. Y. Chan; S. S. Lau *Appl. Surf. Sci.* 2000, 157, 29-34.
[52] D. Mangelinck; J. Y. Dai; J. S. Pan; S. K. Lahiri *Appl. Phys. Lett.* 1999, 75, 1736-1738.
[53] F. M. d' Heurle; P. Gas *J. Mater. Res.* 1986, 1, 205-211.
[54] F. M. d' Heurle; C. S. Peterson; J. E. Baglin; S. J. LaPlaca; C. Y. Wong *J. Appl. Phys.* 1984, 55, 4208-4218.
[55] JCPDS, 38-844, (1988).
[56] JCPDS, 7-251, (1988).
[57] D. Mangelinck; J. Y. Dai; J. S. Pan; S. K. Lahiri *Appl. Phys. Lett.* 1999, 75, 1736-1738.
[58] J. F. Liu; H. B. Chen; J. Y. Feng; J. Zhu *Appl. Phys. Lett.* 2000, 77, 2177-2179.
[59] Semiconductor Industry Association, *International Technology Roadmap for Semiconductors:* 2003 edition
[60] H. Jeon; C. A. Sukow; J. W. Honeycutt; G. A. Rozgonyi; R. J. Nemanich *J. Appl. Phys.* 1992, 71, 4269-4276.
[61] B. I. Boyanov; P. T. Goeller; D. E. Sayers; R. J. Nemanich *J. Appl. Phys.* 1998, 84, 4285-4291.
[62] L. A. Clevenger; R. A. Roy; R. W. Roy; K. L. Saenger; C. Cabral. Jr. *J. Appl. Phys.* 1994, 76, 7874-7881.
[63] A. Appelbaum; R. V. Knoell; S. P. Muraka *J. Appl. Phys.* 1985, 57, 1880-1886.
[64] F. M. d'Heurle; *J. Mater. Res.* 1988, 3, 167-195.
[65] JCPDS 43-989, (1988).
[66] T. G. Finstad *Thin Solid Films* 1978, 51, 411-424.

INDEX

A

absorption, 8, 12, 87, 90, 93, 103, 104, 110, 112, 115, 140, 141
absorption spectrum, 12, 112, 115
acetone, 3, 129
acidity, 91
activation energy, ix, 154, 164, 167, 170, 173, 175
adhesion, 142
adhesive(s), 82, 91
adsorption, 139
aggregates, 109, 110
AIBN, 99
alcohol, 90, 151
algorithm, 7, 13, 15, 19, 28, 29, 31, 32, 77, 78, 79
alloys, ix, 111, 125, 126, 138, 142, 147, 148
aluminum (Al), 16, 106, 112, 147
aluminum surface, 16
ammonia, 96, 106, 107, 109, 114
amorphous phases, 164
anisotropy, 90
annealing, ix, 107, 115, 125, 127, 133, 134, 135, 137, 138, 139, 142, 146, 147, 153, 155, 158, 160, 161, 165, 167, 169, 170, 172, 173
aqueous solutions, 9, 14
asymmetry, 55
atmospheric pressure, 3
atomic force microscope (AFM), 131

B

backscattering, ix, 149, 153, 161
bandgap, 103, 113
basicity, 91
baths, 16
batteries, 100
beams, 22, 49
bending, 12, 20
benzene, 88, 96, 97, 99, 102, 107, 108
biomaterials, 104
biotechnology, 148
birefringence, 8
bismuth, 126, 146, 150
blocks, 155
blood, 148
blood vessels, 148
Boltzmann constant, 132
breakdown, 53, 54, 77

C

calcium, 122
capillarity, 127, 146, 147
carbon, ix, 3, 86, 88, 90, 125, 126, 127, 128, 132, 133, 134, 137, 138, 142, 147, 149, 150, 151
carbon film, 147
carbon nanotubes, 86
carbon tetrachloride, 3
catalyst(s), 91, 104, 111, 112
catalytic activity, 111
cell, 11, 16, 18, 19, 21, 23, 29, 40, 49, 78, 100, 104, 149, 169
CH_2, 101
characterisation, vii
chemical deposition, 126
chemical properties, viii, 85, 113
chemical stability, 3
chemical vapor deposition (CVD), vii
clusters, 139, 141
C-N, 132, 133, 134, 137
coating(s), 91, 95, 111
cobalt, ix, 153, 164
combined effect, 53, 57
composite(s), 86, 91, 104, 112, 114, 116
composition, 126, 133, 138, 142, 147, 170

compounds, viii, 85, 90, 100, 104, 109, 110, 111, 112, 114, 116
concrete, 149
condensation, ix, 125
conductivity, 78, 112, 113
confinement, 147
conversion, 86, 91, 168, 172
cooling process, 2, 24
copper, 104
corrosion, 113
covalent bonding, 113
covering, 64, 69
crystal growth, iv, vii, viii, 1, 2, 3, 4, 5, 8, 9, 10, 11, 12, 13, 14, 15, 18, 19, 20, 21, 22, 24, 27, 29, 31, 43, 44, 45, 46, 49, 50, 58, 60, 62, 63, 64, 67, 68, 70, 75, 76, 77, 79, 80, 81, 82, 83, 85, 86, 143, 144, 146
crystallinity, 110
crystallization, vii, 11, 89, 114, 115, 142, 146

D

data analysis, 11, 15, 22, 23, 30, 40, 76
decay, 62
decomposition, 111, 112
defects, 46, 161
deformation, 19, 26, 40, 45, 49, 70, 73
density, vii, viii, 1, 2, 3, 4, 7, 11, 12, 16, 21, 22, 24, 31, 37, 40, 43, 45, 50, 53, 55, 58, 62, 64, 70, 75, 79, 80, 125, 126, 127, 130, 134, 141, 148, 154, 155
deposition, ix, 4, 19, 31, 43, 45, 46, 49, 50, 62, 64, 68, 69, 70, 77, 125, 127, 129, 130, 132, 139, 142, 150, 151, 152, 153, 161, 165
deposition rate, 70, 165
desorption, 139
diagnostics, 86
diamine, 111
dielectrics, 113
diffusion, 3, 5, 8, 9, 10, 11, 12, 13, 15, 19, 45, 46, 50, 53, 54, 57, 62, 64, 66, 67, 68, 69, 70, 77, 81, 155, 161, 164
diffusivity, 62
displacement, 12, 25, 26, 44, 45
divergence, 4
DMF, 87
drug delivery, 148, 150

E

effect of surfactant, 109
efficiency, 91, 104, 111
elasticity, 139, 141
electric charge, 141, 148
electric conductivity, 148
electrical conductivity, 112, 114
electrode(s), 9, 81, 91, 104, 111, 148
electroluminescence, 100, 147
electrolysis, 10
electrolyte, 10
electron diffraction, 155
electron microscopy, ix, 153, 154, 161, 163
electronic materials, vii
electrons, 135, 141
emission, 86, 91, 103, 110, 115, 127, 129, 131, 132, 140
energy, 16, 32, 37, 62, 86, 91, 127, 128, 130, 131, 139, 141, 142, 146, 149, 150, 160, 164, 167, 169, 170, 173, 174, 175
energy density, 128, 141, 142
energy transfer, 37
engineering, 6, 7, 41, 126
entropy, 173, 174
environment, 4, 12, 85
epitaxial films, 131
epitaxial growth, ix, 153, 161
erosion, 113
ethanol, 3, 87, 91, 94, 96, 115, 129
ethylene, 6, 102, 111
evaporation, ix, 125, 126, 147, 152, 170
excitation, 93, 140, 141
exposure, 8, 132
extrapolation, 15, 29, 30, 31, 32, 33, 77

F

fabrication, 16, 113, 126
fibers, 110
filler, 145
film thickness, ix, 137, 148, 154, 170, 172, 174, 175
films, ix, 111, 125, 127, 131, 142, 153, 154, 155, 159, 160, 161, 163, 164, 165, 167, 168, 170, 172, 174
flame, 91
flatness, 16, 18
fluid, viii, 2, 3, 4, 5, 6, 7, 12, 13, 16, 17, 24, 27, 29, 32, 33, 34, 38, 41, 43, 45, 50, 53, 54, 55, 57, 62, 63, 70, 75, 76, 77, 78, 79, 80, 81, 83
free energy, 164, 167, 173, 174
frequency, 7, 31, 78, 127, 130, 141
friction, 113
Fullerenes, 148
fusion, 2
future, ix, 148, 149, 150, 154, 160, 164

G

galaxy, 149
gallium phosphide (GaP), 126
gel, 10, 12, 81, 82, 93, 95
gene, 148
Germany, 83, 122
glasses, vii
glucose, 87
glycine, 3
glycol, 88, 89, 106, 111
gold, 126
grains, 115, 132, 146, 159, 160
graphite, ix, 125, 127, 131, 132, 134, 135, 136, 137, 142, 145, 146, 151
gravitational field, 62
gravity, 22, 54, 62, 63, 78, 82

H

hardness, 104, 113, 129
heat, viii, 2, 6, 27, 37, 78, 80, 141, 142
heat transfer, viii, 2, 6, 27, 37, 80
helium, 6, 22
heterogeneity, 138
hexane, 3, 94
High Resolution Transmission Electron Microscopy (HRTEM), 154, 161, 163, 167, 168, 174, 175
holes, 146
hydrazine, 96
hydrogen, viii, 2, 3, 86, 111
hydroxide, 90, 91

I

impurities, 8, 12, 93, 154
India, 1, 78
indium, ix, 125, 126, 127, 128, 131, 132, 133, 134, 137, 138, 139, 141, 142, 144, 145, 146
initial state, 147
insulation, 18
integrated circuits, 160
interface, 8, 10, 12, 40, 46, 82, 154, 158, 160, 163, 164, 167, 168, 173
interface energy, 160, 164
ionization, 139, 141
ions, 91, 113, 139, 141, 142
IR, 100
IR spectroscopy, 100
irradiation, 12, 134, 136, 141, 159, 169
isotherms, 6, 32, 33

K

KBr, 129, 130, 134, 135
kinetics, vii, 8, 9, 11, 12, 14, 154
KOH, 87, 95, 105

L

laser ablation, 126, 127, 132, 139, 140, 141, 142, 151
lasers, 2, 6, 100
lattice parameters, 155, 158, 160, 168
layering, 46
light, 4, 6, 8, 9, 12, 15, 22, 23, 24, 25, 26, 27, 29, 34, 37, 40, 41, 44, 45, 46, 48, 49, 53, 54, 58, 65, 67, 69, 72, 75, 78, 100, 103, 113, 126, 141, 148, 149
light beam, 23, 24, 25, 26, 27, 29, 34, 45, 49, 67, 72
light emitting diodes, 100
light scattering, 6
liquid phase, 142, 144
lithium, 3, 100
luminescence, 91, 100, 104, 113
luminescent devices, 113
lysozyme, 2, 12, 82

M

macromolecules, 139
magnetic field(s), 3, 12, 13, 82
magnetic properties, 111
magnetism, 96, 105, 106, 148
mass, 3, 5, 7, 27, 30, 50, 62, 79, 132, 139, 141
material surface, 139
mechanical properties, 149
medicine, 148
melt(s), vii, ix, 1, 4, 125, 139, 142, 146
melting, viii, 125, 139, 142, 144, 146, 147
membrane(s), 104, 146
metal hydroxides, 90
metals, vii, 100, 113, 132, 137, 141, 147
methanol, 3, 87, 89, 91, 93, 94, 96, 112
microscope, 9, 10, 81, 130, 131, 149
mixing, 54, 80, 173, 174
mobility, 113
molecular mass, 4
molecular structure, 12
monomer, 101
morphology, viii, 3, 10, 11, 12, 15, 68, 75, 85, 89, 90, 92, 93, 101, 102, 111, 127, 131, 168

N

NaCl, 134, 146, 152
nanobelts, 86, 87, 90, 91, 95, 96, 116
nanocrystals, viii, 85, 89, 90, 91, 92, 93, 94, 99, 101, 103, 109, 113, 114
nanodots, viii, 85, 86, 87, 89, 90, 91, 94, 116
nanomachines, 148
nanomaterials, viii, 85, 90, 91, 100
nanoparticles, 86, 87, 90, 92, 93, 94, 95, 96, 97, 98, 99, 100, 102, 103, 104, 105, 106, 107, 108, 109, 110, 111, 112, 113, 114, 115, 116
nanoribbon(s), 97, 98, 100, 101, 102, 103, 116
nanorods, viii, 85, 86, 87, 90, 91, 94, 95, 96, 97, 98, 99, 100, 101, 102, 103, 104, 105, 107, 109, 110, 111, 112, 113, 114, 115, 116
nanostructure(s), viii, 85, 86, 87, 88, 89, 90, 91, 94, 95, 96, 97, 98, 99, 100, 101, 102, 103, 105, 106, 107, 108, 109, 110, 111, 112, 113, 114, 115, 116, 126, 142
nanotube(s), viii, 85, 86, 87, 88, 89, 90, 91, 93, 94, 95, 96, 98, 101, 104, 106, 111, 116, 126, 127, 149, 150
nanowires, viii, 85, 86, 87, 89, 90, 91, 92, 93, 94, 96, 97, 98, 99, 100, 101, 102, 105, 108, 113, 115, 116, 126, 147
Nd, 127, 133, 135, 141, 147
nickel, ix, 125, 127, 128, 129, 131, 132, 133, 134, 137, 138, 139, 141, 142, 144, 145, 146, 164
nitride, 113, 126
non-stationary, 6
nucleation, vii, ix, 4, 18, 146, 154, 158, 160, 164, 167, 170, 173, 174, 175
nuclei, 164, 173

O

optical fiber, 148
optical properties, 93, 111, 112
optoelectronic devices, 91, 100, 113
organic solvents, 92
organs, 148
orientation, 3, 9, 21, 50, 53, 75, 77, 154, 155, 159, 161, 167
oxides, viii, 85, 86, 90, 91, 92, 114, 116

P

particles, 12, 21, 43, 50, 53, 62, 76, 85, 86, 89, 90, 91, 92, 101, 102, 109, 112, 115, 127, 130, 132, 137, 138, 139, 140, 141, 142, 146, 151
pH, viii, 85, 90, 92, 94, 109, 111, 112
phase boundaries, 173
phase conjugation, 9
phase diagram, 142, 143, 144
phase transformation, 168
photoconductivity, 104
photoluminescence (PL), 90, 93, 100, 103, 110, 115
photon(s), 140, 141, 148, 149
physical chemistry, 86
physical mechanisms, 43
physical properties, 86
plasma, 127, 129, 132, 140, 141, 150
polarizability, 26, 79
polarization, 8, 9
pollution, 142
poly(vinylpyrrolidone) (PVP), 87, 89, 96
polyacrylamide, 108
polyethylene, 106, 111
polymer(s), vii, 99, 111
polyvinylalcohol [PVA], 87, 90, 97, 101, 109
porosity, 146
propagation, 16, 24, 25, 40, 49, 78
properties, vii, 1, 4, 6, 12, 40, 63, 80, 86, 89, 90, 91, 93, 94, 95, 96, 103, 104, 109, 110, 111, 112, 113, 115, 116, 127, 144, 148
protective coating, 113
protein(s), 2, 3, 8, 12, 46, 80, 82, 104, 139
protein crystallization, 82
purification, vii

Q

quality, vii, ix, 1, 2, 3, 4, 10, 11, 12, 13, 14, 15, 19, 43, 46, 49, 54, 55, 58, 62, 63, 70, 75, 76, 114, 148, 153, 160, 161, 163, 174
quantization, 113, 126
quantum confinement, 93, 113
quantum dot(s), 126, 148, 150

R

radiation, 104
radiation damage, 104
Raman spectra, 110, 150
Raman spectroscopy, 131, 132, 137
reactant(s), 89, 91, 92, 100, 109, 114
reaction temperature, 86, 89, 91, 92, 102, 103, 109, 110, 111, 112, 114, 116
reaction time, viii, 85, 86, 89, 91, 92, 102, 103, 110, 111, 112, 114, 116
red shift, 103, 110
reduction, 4, 38, 41, 49, 50, 75, 113, 154, 167, 169, 170, 174, 175

refractive index, vii, 1, 2, 4, 6, 8, 9, 10, 11, 13, 14, 24, 25, 26, 27, 40, 45, 64, 75, 82, 104
refractive index variation, 7
research, ix, 6, 43, 86, 149, 153, 154, 160
research and development, 160
resistance, ix, 100, 113, 153, 154, 155, 158, 160, 164, 167, 170, 172
room temperature, 16, 93, 113, 129, 130, 132, 133, 161
rotations, 50, 127
RT, 130, 133, 139, 142

S

scanning probe microscopy, 147
scattering, 24, 86, 146
self-organization, 126
SEM micrographs, 133, 137
semiconductors, vii, 86, 91, 100, 113, 126, 127
sensitivity, viii, 2, 8, 9, 10, 11, 25, 40, 49
sensors, 91, 100
shade, 69
shape, viii, 10, 35, 37, 38, 45, 53, 68, 70, 75, 85, 89, 90, 91, 92, 93, 100, 102, 103, 104, 109, 110, 111, 114, 115, 116, 126, 138, 148, 173
silica, 94
silicon, 22, 126, 154, 174
silver, 10, 22, 87, 89
SiO2, 90, 106, 111, 112
solar cells, 91
solid phase, ix, 144, 153, 161, 163, 164, 172
solid solutions, 113
solid state, ix, 153, 154, 161
solid surfaces, 141
solubility, 3, 12, 21
solution, vii, ix, 1, 2, 3, 4, 5, 7, 8, 9, 10, 11, 12, 13, 14, 15, 18, 19, 20, 21, 24, 25, 26, 27, 28, 29, 30, 31, 32, 42, 43, 44, 45, 46, 47, 48, 49, 50, 51, 52, 53, 54, 55, 57, 58, 59, 60, 61, 62, 63, 64, 65, 66, 67, 68, 69, 70, 71, 72, 73, 74, 75, 76, 77, 78, 79, 80, 81, 82, 85, 86, 89, 90, 91, 104, 110, 125, 127, 134, 143, 144, 146, 168, 169, 170, 173, 174
solvent(s), viii, 3, 12, 85, 89, 90, 91, 93, 100, 102, 107, 109, 114, 169
species, 141, 142
spectroscopy, 129, 131, 149, 151
spherulites, 10
spin, 150
strength, viii, 3, 7, 9, 13, 15, 16, 19, 42, 49, 50, 53, 55, 62, 63, 67, 77, 125, 126
substrates, ix, 129, 131, 134, 153, 155, 158, 160, 161, 163, 164
sugar, 10
superconductor, 100
superlattice, 150
surface area, 9
surface tension, 3, 139
surfactants, 89, 90, 91, 96, 100, 109, 111, 114
symmetry, 3, 11, 13, 15, 45, 46, 53, 54, 55, 58, 62, 63, 67, 68, 70, 72, 73, 74, 76, 77
synthesis, 126

T

temperature, viii, ix, 2, 3, 4, 5, 6, 7, 8, 9, 10, 12, 13, 14, 16, 17, 18, 19, 21, 22, 24, 26, 27, 32, 33, 34, 37, 38, 40, 41, 43, 45, 50, 54, 64, 66, 75, 77, 78, 79, 80, 82, 85, 87, 88, 89, 92, 94, 95, 96, 97, 98, 99, 102, 103, 104, 105, 106, 107, 108, 110, 111, 112, 113, 114, 125, 129, 132, 139, 141, 142, 143, 144, 153, 154, 155, 158, 159, 160, 161, 164, 167, 168, 170, 172, 173, 174
temperature annealing, 164
thermal energy, 141
thermal properties, 34
thermal stability, ix, 153, 154, 160, 164, 167, 168, 170, 172, 175
thermogravimetric analysis (TGA), 97, 98, 99, 101
thin film(s), vii, ix, 125, 126, 127, 129, 130, 131, 132, 133, 134, 135, 136, 137, 138, 139, 140, 142, 144, 145, 146, 150, 151, 152, 154, 160
time resolution, 9, 132
titanium, 106, 158
titanium isopropoxide, 106
toluene, 94, 115
toxicity, 3
transformation, 9, 154, 155, 167
transition metal, 90
transparency, 15, 21, 54, 55, 58, 70, 113
transport, vii, 3, 4, 9, 12, 32, 46, 50, 54, 55, 58, 64, 66, 68, 70, 77, 83
treatment, 28, 93, 103, 110, 115, 116

U

urea, 3, 94
USA, 85
UV, 12, 87, 90, 91, 97, 110, 112, 113, 141
UV light, 12, 112

V

vacuum, 129, 141, 142, 145, 155, 160, 161, 165, 170
validation, 5, 15, 32, 82
vapor-liquid-solid (VLS), 126, 127, 145, 146, 152

viscosity, 3, 62, 79

W

wall temperature, 27
water, viii, 2, 3, 6, 10, 13, 14, 16, 17, 18, 19, 21, 27, 40, 41, 42, 75, 76, 89, 90, 92, 110, 115, 130, 134, 135, 155
wavelengths, 132
wear, 100, 113
wires, 133
wood, 124

X

xenon, 6
X-ray diffraction (XRD), 114, 131, 155, 156, 157, 159, 161, 162, 165, 166, 167, 168, 169, 170, 171, 172

Z

ZnO, 90, 91, 92, 94, 106, 111, 112
ZnO nanorods, 91, 112
ZnO nanostructures, 92